INTERNAL REFLECTION AND ATR SPECTROSCOPY

CHEMICAL ANALYSIS

A SERIES OF MONOGRAPHS ON ANALYTICAL CHEMISTRY
AND ITS APPLICATIONS

Series Editor
MARK F. VITHA

Volume 176

A complete list of the titles in this series appears at the end of this volume.

INTERNAL REFLECTION AND ATR SPECTROSCOPY

Milan Milosevic

MeV Technologies LLC

WILEY

A JOHN WILEY & SONS, INC., PUBLICATION

Published by John Wiley & Sons, Inc., Hoboken, New Jersey
Published simultaneously in Canada

Library of Congress Cataloging-in-Publication Data:

Milosevic, Milan, 1955–
Internal reflection and ATR spectroscopy / Milan Milosevic.
p. cm.
Includes index.
ISBN 978-0-470-27832-1 (cloth)
1. Internal reflection spectroscopy. 2. Absorption spectra. I. Title.
QC454.I5M55 2012
543'.59–dc23

2011046733

Printed in the United States of America

ISBN: 9780470278321

10 9 8 7 6 5 4 3 2 1

To my wife and son

CONTENTS

PREFACE

This book attempts to provide a bridge between the topics of electromagnetic theory and spectroscopy, and in particular attenuated total reflection (ATR) spectroscopy. Electromagnetic theory is typically addressed in physics textbooks dealing with electromagnetic wave propagation in media and the reflection and refraction of the electromagnetic wave at interfaces between different media. A typical physics textbook derives Fresnel equations, mentions Brewster's angle, and total internal reflection as curiosities, and swiftly moves on to other phenomena. As far as physics textbooks are concerned, the general problem of reflection and transmission is completely solved. No new physics beyond this solution is anticipated, and as far as physics is concerned, it is dead as a research topic.

Analytical chemists, who are the primary users of spectroscopy, see the entire topic of reflection and transmission spectroscopy as a tool for finding answers about their samples. They typically concern themselves with the interpretation of spectroscopic experiments and, for their purposes, it is sufficient to assume that the absorption peaks seen in the spectra are the result of light at those frequencies being absorbed by the sample and that the absorption strength is proportional to the concentration of the absorbing substance. Analytical chemists pay only minimal attention to the physics of spectroscopic experiments. As tools, those experiments are employed to provide answers to the questions about samples that are analytical chemists' primary interest. Those cases in which analytical chemists pay any attention to the physics that underlies spectroscopy are when they encounter the so-called optical effects that interfere with the usual assumptions about the spectra such as linearity or where the features in the spectra are not consequences of light absorption by samples, but of some other phenomenon.

There is not much precedence for a book on this subject. A typical subject covered by textbooks can rely on a collection of topics that has evolved over time and distilled through many textbooks written by many authors as comprising the most appropriate set of topics for the subject. For a book attempting to cover an essentially uncovered subject, this accumulated wisdom is lacking. It is not a priori clear which topics should be addressed explicitly, which should be assumed to be in the background knowledge of the reader, and which topics should be left out as being beyond the scope of the book. Thus, the choice of topics addressed in this book reflects the preferences and fascinations of the author. In addition, the process of writing the book forced its own selection and order of topics as some were needed as prerequisites for others. Another phenomenon stepped in as well. The systematic treatment of the subject undertaken here brought up its own surprises. In following the logic of the narrative, we stumbled on some previously undescribed phenomena and effects. There was then no way of avoiding covering these phenomena, although they are not in a current vocabulary of the field.

In addition, I used the experience derived from thousands and thousands of discussions with researchers all around the world who needed specialized devices for spectroscopic measurements. Different spectroscopic measurements span a huge range of various applications of optical spectroscopy and require the choice of a spectroscopic technique most suited to a particular experiment. It was thus important to anticipate different benefits a spectroscopic technique and a particular experimental setup would have, so the most likely to succeed approach can be taken. The accumulated residue of all these discussions served as a compass for the choice of topics covered and for the assumed typical background possessed by a potential reader.

I assumed that the reader is somewhat familiar with the basic theory of classical electrodynamics and with basic calculus. Some of the topics covered could have been left out of the book as already familiar to the reader but were included for completeness and to provide coverage of the subject with unified notation and within a single system of units.

The main character in this book is the evanescent wave. I covered the evanescent wave in many different ways, both to help the reader understand it as a physical phenomenon and to help the reader understand ATR spectroscopy, which is based on it. I initially described evanescent wave as it is described in a typical textbook on electrodynamics or physical optics. This description suffices for most of what we want to understand about the evanescent wave and ATR spectroscopy. It took almost the entire book until this conventional understanding ran into a proverbial brick wall and the received wisdom needed to be reevaluated. This happened as I covered the ATR spectroscopy of powders. It happened in the course of writing the chapter and was forced out by the need to provide clear understanding of the ATR spectroscopy of powdered media. Some of the dead ends in this effort were left in the text to help the reader appreciate the difficulties the conventional understanding was not able to resolve.

Although the evanescent wave is the main character in this book, the unsung hero that emerges from the narrative is the classical electromagnetic theory and its description of reflection, transmission, and wave propagation through media. This description remained meaningful as we pushed it to incorporate more and more phenomena that we wanted to understand. The formalism readily incorporated the description of absorbing media, the extension of internal reflection into the regime of supercritical angle of incidence, and the extension into metal optics.

In the text, the words light and electromagnetic radiation are used interchangeably. Light is any electromagnetic radiation that is manipulated by usual optical means such as lenses and mirrors. Therefore, light is any electromagnetic radiation, from far infrared (IR) (wavelength about 0.1 mm) to deep UV (wavelength about 0.1 μm), not just visible light.

A short discussion of certain aspects of actual ATR experiments was incorporated since these are of interest to the intended readers and generally belong under the topic of "optical effects." These effects fall into the gap between the subjects of physical optics and analytical chemistry and are thus rarely discussed. However, these effects, as demonstrated in the text, can cause a nonlinear response of the absorbance transform of ATR spectra even if it is assumed that for ATR, in the absence of these effects, the absorbance transformed spectrum would exhibit a linear response.

Explicit effort was made to avoid quantum mechanics in presenting the concepts and results. For the most part, it is possible to describe the subject within the formalism of classical physics. It is, of course, not possible to avoid quantum mechanics when trying to further elaborate on most of the concepts discussed in the text, but it is satisfying not to have to drag in the unintuitive murkiness of quantum physics in order to understand the physical mechanisms underlying the phenomena utilized in spectroscopy.

In discussing how the various spectral features relate to a sample or to an experimental setup, we resorted to the numerical approach. All spectra shown in the book are numerically simulated using an early prototype of the SimSpec™ software. We found numerical simulations preferable to actual spectra since, in a numerically simulated spectrum, everything is under explicit control, so it is easier to sort out which parameter, whether sample or experimental setup related, is responsible for a particular spectral feature. For instance, it would be much harder to deposit an ultrathin film of a known thickness than to simply type in the appropriate number into the software. Also, experimental parameters such as the angle of incidence or polarization are not easy to set accurately in an actual experiment, while they are set with perfect accuracy in a numerical simulation. Numerically calculated spectra are easy to study. The mathematical expressions from which these spectra are calculated involve absolute values of complex numbers. That makes them difficult to scrutinize analytically, so their behavior is hard to infer directly from their form. Thus, it should not be surprising that careful investigations of these expressions can turn new and unexpected results despite the fact that these expressions have been around for a long time.

The efforts that my wife Violet put in helping me complete this book were instrumental to the project coming to completion. She carefully read the manuscript, made many helpful comments and suggestions, and provided support, encouragement, and inspiration.

MILAN MILOSEVIC
MeV Technologies LLC

Westport, Connecticut
October 2011

1 Introduction to Spectroscopy

1.1 HISTORY

Spectroscopy got its start with Newton's observation that the white light from the sun can be separated into different colors using a prism. This observation became not only foundational to the theory of light but also to the understanding of human color vision. Color vision is a crude form of spectroscopy. We can often guess on the nature of a material by its color. We can only perceive three primary colors. That makes the human eye a very crude spectrometer. However, even at this crude level, it was apparently advantageous to us, during evolutionary history, to trade higher image resolution for color vision.

In the times of Newton, the hurdle to overcome was to understand the nature of light. Newton himself proposed the so-called corpuscular theory of light with which he was able to explain all then known properties of light (i.e., the propagation of light in a straight line and the laws of reflection and refraction). Newton proposed that light consists of tiny particles (corpuscles). Huygens, a contemporary of Newton, proposed the so-called wave theory of light, in which light is a wave phenomenon like a wave on the surface of water. Using his theory, he too was able to explain all the known properties of light. To make their respective theories work, Newton and Huygens made opposite assumptions about the speed of light in optically transparent media such as water or glass. Newton needed light to go faster through such media for his theory to work. Huygens needed it to go slower.

However, what settled the dispute was not the measurement of the speed of light in transparent media; it was the observation that light can form interference patterns. Only waves can form interference patterns and that settled the dispute. Huygens won. Later, measurements of the speed of light in denser media such as water confirmed Huygens's assumption. Light was a wave phenomenon.

Internal Reflection and ATR Spectroscopy, First Edition. Milan Milosevic.

Different colors of light correspond to different wavelengths. In studying light dispersion by a glass prism, Herschel noticed that there is an invisible component of solar radiation next to red light. Thus, infrared (IR) light was discovered. Later, it was discovered that there is also an invisible component of solar radiation next to violet that was named ultraviolet (UV).

In the early nineteenth century, Fraunhofer noticed a curious phenomenon. By using a prism to disperse solar radiation, he observed tiny black lines superimposed over a continuous rainbow-colored solar spectrum. Some wavelengths in the continuous spectrum of sunlight were missing. The dark lines, later named Fraunhofer lines, were present whether he used one type of glass prism or another. The appearance of the lines was mysterious, but Fraunhofer could find them not only in the solar spectrum; he was also able to observe them in the spectra of distant stars. Following the work of Thomas Young with interference of light, Fraunhofer developed diffraction grating as a means to disperse light in a more effective way than with glass prisms. The gratings enabled spectroscopy with a much higher resolution than prisms and enabled the direct measurement of the wavelengths of light.

Fraunhofer died without knowing what caused his dark lines. It was Kirchoff, working some 30 years after Fraunhofer's death, who realized that each element or compound is associated with its own unique set of spectral lines. This was the official birth of spectroscopy as a scientific discipline. The implications of this observation were extremely important. For instance, it told us that distant stars were made from the same elements found here on Earth. We could not know that any other way with any certainty. This is a nontrivial finding about the nature of the universe. There is no a priori reason why the distant worlds, many light-years from Earth, would not be made from an entirely new type of matter, entirely different from what we are familiar with here on Earth.

Intrigued by these mysterious lines and their association with different elements, many researchers started studying spectra of flames and other light sources. It was discovered that, when heated, the atoms emit bright lines. Soon it was realized that these bright lines match some of the dark lines found in the solar spectrum. Associating lines with particular elements became the primary aim of the new science of spectroscopy. Soon people would talk about sodium lines, mercury lines, and so on.

It was also realized that the dark lines were due to absorption of light by the elements that would, when heated, emit those same lines. Soon, it became possible to analyze a spectrum of a mix of elements by sorting out the lines due to each element, that is, to analyze a mixture for its constituents. Furthermore, by observing the relative intensities of lines due to each element, it became possible to estimate the relative abundance of different elements in a mixture. This was now already true spectroscopy.

Early on, spectroscopists realized that they could substitute a photographic plate for their eyes and that they could photograph a spectrum. The spectrograph represented a permanent record of a spectrum and could be subsequently analyzed in great detail. Different spectra could be compared. Employing long exposure times allowed recording spectra from very faint sources otherwise too weak to be observed by the eye. The use of photography also extended the spectral range of spectroscopy from visible to UV and, to a more limited extent, to the IR spectral regions.

By improving the spectroscopic equipment and increasing the resolution of the spectroscopic measurement, spectroscopists soon realized that many single lines seen through the early spectroscopes were not really single and that sometimes, under high spectral resolution, a much finer structure would be revealed. They found single lines resolving into doublets, triplets, quadruplets, and so on. By the dawn of the twentieth century, a great amount of very detailed spectral information was amassed. The experimental precision with which these spectral measurements were pursued seems almost fanatical, but what propelled it was the constant stream of discoveries that accompanied it. For instance, it was discovered that some prominent lines in the sun's spectrum could not be matched by anything known on Earth. They attributed it to a new element that they aptly named helium after the Greek sun god Helios. Soon thereafter, helium was discovered on Earth.

However, the abundance of information generated by spectroscopy was contrasted with the total lack of understanding of how the spectra themselves are generated. People knew that light is a wave phenomenon similar to sound. The sound generated by a taut string consists of a set of characteristic frequencies. A string with a different tension or of a different length produces a sound of a different frequency. This would make it plausible that different elements would produce different sets of light frequencies. Even the fact that a taut string could be resonantly excited into vibrations by a sound of the same frequency that it would sound if struck was seen analogous to why cold atoms would absorb the same frequencies of light that they would emit when heated.

While not in itself surprising, the existence of these characteristic frequencies associated with different elements was totally stomping the scientists when they tried to understand them based on the available physical theories known collectively as classical physics. Soon, it became obvious that classical physics could not explain the observed spectra. A revolutionary new theory called quantum mechanics had to be developed to provide the explanation. The explanation, however perfect, came with an uneasy requirement to abandon common sense and to proceed into the unintuitive and forbidding world of quantum mechanics following mathematics where intuition fails.

After first providing a spectacular confirmation that the universe is filled with the same atoms and compounds that we find here on Earth, spectroscopy provided another spectacular result. Measuring spectra of distant nebulae in the first half of the twentieth century, Hubble discovered that the spectral lines of elements and compounds from those distant nebulae are shifted from their terrestrial positions toward lower frequencies (referred to as redshift since red light is the visible light with the lowest frequency). This was a puzzling discovery.

The explanation that was eventually accepted is that those distant nebulae recede from us in all directions with high speeds. The recession at high speeds shifts frequencies through what is known as Doppler effect. The effect is commonly observed when a whistling train passes by. The pitch of the whistle is higher while the train is approaching and it suddenly turns lower as the train passes by and starts moving away.

By studying how the redshift correlates with the distance from Earth, Hubble found that the farther away a nebula (today referred to as galaxy) is, the larger the redshift. This finding stood in a distinctly anti-Copernican spirit; that is, that Earth has no special place in the universe, but it was soon realized that the same is true for every point in the universe. The universe is expanding from every point in every direction. The most significant implication of that observation is that, by playing backward the movie of the expansion, we find that the entire universe came into existence in a huge "bang" some 13.7 billion years ago. Spectroscopy thus ushered the age of the big bang cosmology.

Ultraviolet and visible spectroscopy could progress by using photographic techniques to record spectra. However, IR spectroscopy could not since the sensitivity of the photographic plates to IR light greatly diminishes for the wavelengths longer than those of red light. Thus, to pursue IR spectroscopy, a new way of light detection had to be employed. Early in the twentieth century, Coblentz developed and used a thermopile detector to push spectroscopy far into the IR. He used rock salt prisms (as opposed to glass prisms, which are opaque to longer wavelengths) to disperse light and placed a thermopile detector to detect the IR light of a selected wavelength. The thermopile detector consists of a thermocouple and a voltmeter. In this way, he could read the voltage produced by the thermocouple when heated by IR light at the selected wavelength. A graph connecting the thermocouple response versus wavelength is called the spectrum. In this way, working painstakingly, Coblentz collected a large number of very high quality spectra of a great variety of compounds. The amazing specificity of the IR spectra of different compounds rivaled the specificity of atomic spectra. Coblentz soon realized that different functional groups are characterized by specific absorption peaks. Thus, for instance, the presence of a C-H group in a molecule was revealed by characteristic spectral absorptions. The same is true for the O-H and other groups. Thus, IR spectroscopy became a great tool not only to identify an unknown substance by comparing its spectra to the spectra of known compounds, but it allowed the molecular structure to be inferred from the information contained in its IR spectrum directly, just based on the known absorption bands associated with various functional groups. Coblentz's work established modern IR spectroscopy. The period following World War II saw the introduction of the first commercial IR spectrometers. They had motorized wavelength scanning and produced plots of spectra using chart plotters. Spectral collection was transformed from painstaking work taking hours to routine scans taking a few minutes each.

Early spectroscopy was almost exclusively transmission spectroscopy. Some reflection spectroscopy was pursued within the research community interested in optical properties of metals and within the mirror making community, which needed a way to properly characterize its offerings. Reflection spectroscopy was also of interest for crystallography. However, it was not until the advent of attenuated total reflection (ATR) spectroscopy that a spectroscopic technique other than transmission spectroscopy was routinely used. Attenuated total reflection spectroscopy is an unlikely spectroscopic technique. It is based on the phenomenon of total internal reflection that has been known for a long time and has even been discussed by Newton in his *Opticks*. During total internal reflection, a special type of electromagnetic wave called the evanescent wave is formed on the other side of the reflecting interface. An absorbing material brought into contact with the totally reflecting interface absorbs some of the intensity of the evanescent wave and the intensity of the reflected light is thus attenuated with respect to the incoming intensity—hence, the name ATR spectroscopy. In many respects, ATR spectra resemble transmission spectra. That helped ease the acceptance of ATR spectroscopy by the IR spectroscopy community. Attenuated total reflection spectroscopy was proposed in 1959–1960 independently by Harrick and Fahrenfort.

Fahrenfort approached ATR from the single reflection side. Harrick approached it from the multiple reflection side.

Standard commercial spectrometers are built for transmission measurements so reflection measurements could not be done with these spectrometers. Researchers in the field had to design and build their own devices that allowed them to use ATR spectroscopy in their existing spectrometers. Some of the first commercial ATR accessories for use in the existing commercial spectrometers were made by Wilks, enabling anybody with an IR spectrometer to use ATR spectroscopy. Multiple reflection ATR spectroscopy initially took hold. Multiple internal reflection elements became available in a number of standardized sizes and were offered by multiple vendors. The popular ATR materials included germanium, KRS-5 (thalium bromoiodide), silicon, ZnSe, and ZnS. The initial rationale for using multiple reflection ATR was that the absorbance produced in ATR spectroscopy is very weak and that the only practical way to reach a good signal/noise (S/N) of a measurement was to use multiple reflections.

The second half of 1980s saw the wholesale replacement of older grating-based spectrometers with new Fourier transform infrared (FTIR) spectrometers. This replacement drastically increased spectrometer performance, enabling much higher S/N measurements. This rekindled the interest in single reflection ATR spectroscopy. Starting in the early 1990s, the interest in single reflection ATR spectroscopy got a further boost from the interest in ATR microsampling. Microsampling single reflection ATR spectroscopy soon became the dominant spectroscopic technique in IR spectroscopy. This trend was further stimulated by the introduction of diamond as the ATR material. The mechanical strength, abrasion resistance, chemical inertness, and great optical characteristics of diamond propelled diamond single reflection ATR spectroscopy into a nearly universal spectroscopic technique. A broad range of sample types, from hard solids to liquids to powders, can all be readily analyzed using a single diamond ATR accessory.

1.2 DEFINITION OF TRANSMITTANCE AND REFLECTANCE

Transmittance and reflectance are quantities measured in a spectrometric experiment. A spectrometric experiment consists of a light source, a discriminator to separate contributions from different frequencies (or wavelengths) of light, and a detector. The setup may also include suitable optics to guide light from the source to the detector.

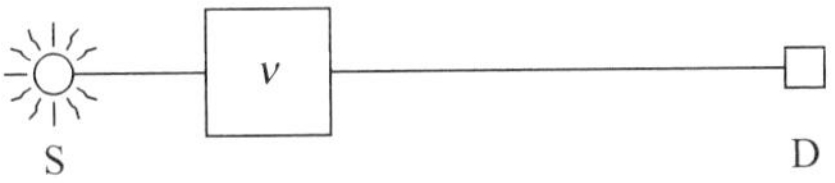

Figure 1.1 Schematic representation of a spectrometric experiment. Source S sends light of all frequencies through the monochromator, which selects only light of frequency ν to proceed to detector D.

The source emits light of different frequencies. The light then goes through the discriminator. The simplest type of discriminator is a prism that splits light into different frequencies (or wavelengths) and allows one frequency at a time to reach the detector. This type of discriminator is called a monochromator. One can also use an optical grating instead of a prism to split white light into its constituent frequencies. An array of detectors could be used instead of a single detector, and all the frequencies (wavelengths) of light could be detected simultaneously. Yet another way to discriminate different frequencies is to modulate each frequency of light with a different frequency as it is done with a scanning Michelson interferometer, to record the signal as a function of the modulation parameter, and subsequently, to transform mathematically the recorded signal into a spectrum. A spectrum of a source is a graph that shows how much power is emitted by the source at different frequencies (wavelengths) of light. Acquiring a spectrum is the goal of a spectrometric experiment.

A spectrometer, shown schematically in Figure 1.1, is a device specifically developed to collect spectra. Every spectrometer consists of a wavelength discriminator and a detector. Most also incorporate a source of light. In addition, there is a lot of control electronics. The light source is usually a hot body that emits light such as a bulb, but it could also be an LED, a laser, plasma, and so on. Or the source could be external to the spectrometer as it is for spectrometers attached to astronomical telescopes used to acquire spectra of distant stars. If the sample itself is the source of light that is analyzed by the spectrometric experiment, then the resulting spectrum is called emission spectrum. As we are interested in ATR spectroscopy, we are not going to cover emission spectroscopy here. We always assume that a source is integral to the spectrometer and that the spectra collected are those of the integral source modified by the presence of the sample. It is important that the source emits light throughout the spectral region of interest and that the light output is stable; that is, it does not fluctuate with time.

For the sake of conceptual clarity, let us assume that for frequency discrimination, we are using a monochromator, a black box with input and output ports for light and a knob with a dial that allows the selection of the output wavelength. On one side comes in white light from the source, on the other side goes out light of a selected frequency (or wavelength).

A detector is a device that converts the power of incoming light into voltage. The voltage is then digitized; that is, the analog electrical signal is converted into a number. The number representing light intensity is then stored for further manipulation.

Now we can see the process of collecting a spectrum as turning the knob on the monochromator to a starting frequency and recording the intensity, moving the monochromator to the next frequency and recording the intensity and so forth until the spectral region of interest has been covered. The result is a spectrum, that is, a series of pairs of numbers in the form (frequency, intensity). We can thus collect a spectrum of the spectrometer's source $I_0(\nu)$. Imagine that we now put a sample into the beam and collect the spectrum of light from the spectrometer's source that has transmitted through the sample $I_S(\nu)$. The ratio

$$T(\nu) = \frac{I_S(\nu)}{I_0(\nu)} \tag{1.1}$$

is called the transmittance of the sample. Note an important characteristic of the thus defined transmittance. The spectrum $I(\nu)$ is the detector's response associated with the intensity of light of frequency ν. The voltage the detector puts out depends on a number of factors such as source intensity, detector area, sensitivity, spectral response, amplifier gain, and so on. Usually, a spectrum is recorded in arbitrary units. However, it is important for the detector signal to be proportional to light intensity. If it is, we say that the detector has a linear response. Thus, as we take the ratio (Eq. 1.1), all other factors cancel out and the result is the ratio of the light intensities with the sample and without the sample in the beam. All the factors related to the source, monochromator, detector, preamplifier, and so on, cancel out. The transmittance thus recorded is entirely a property of the sample itself. This is of fundamental importance. What it says is that the same transmittance of the sample would have been recorded if we had a different source instead of the original one, or a different monochromator, or a different detector. The spectrum measured, at least in principle, depends only on the sample and not the spectrometer used for the measurement. In practice, of course, this is not automatically the case, and a lot of ingenuity has to be expended on the design and manufacture of spectrometers to ensure that this is indeed true. We will come back to investigate this topic in more detail.

From definition (Eq. 1.1), we conclude that the transmittance of a sample is always a quantity smaller than one (i.e., 100%). This is so because, as light impinges on the sample, it partially reflects back from the sample, or the portion of light intensity that entered the sample can be absorbed by the sample. Both processes reduce the intensity of light on the detector as compared to the light intensity that would reach the detector had the sample not been in the beam. Therefore, the numerator in Equation 1.1 is always smaller than the denominator, and the transmittance is always less than, or at most, equal to one.

Instead of being interested in light that transmits through the sample, we could be interested in light $I_R(\nu)$ that reflects from the sample. In this case, we define the reflectance of a sample in analogy with the definition of the transmittance (Eq. 1.1); that is,

$$R(\nu) = \frac{I_R(\nu)}{I_0(\nu)}. \tag{1.2}$$

Again, we expect I_R to be always smaller than I_0. This is so because only a portion of light that was incoming onto the sample could have been reflected from it. Thus, reflectance is always less than one. It is also independent of the source of light used. If the sample does not absorb light of a particular frequency, the incoming light could be either transmitted through the sample or reflected from it. This means that for a nonabsorbing sample,

$$R(\nu) + T(\nu) = 1.$$

For an absorbing sample, we can define the quantity $A_b(\nu) = I_A/I_0$, where I_A is the intensity of light absorbed by the sample. This quantity is not called absorbance, although this would seem natural in analogy with the definitions of transmittance and reflectance. The reason for this is that the term absorbance has been historically attached to another quantity. Nevertheless, the quantity A_b completes the selection of quantities describing the interaction of a sample with light. Incident light can either be reflected from the sample, transmitted through it, or become absorbed by the sample; that is,

$$I_0(\nu) = I_A(\nu) + I_R(\nu) + I_T(\nu).$$

Consequently,

$$R(\nu) + T(\nu) + A_b(\nu) = 1$$

is always true for any sample. We cannot measure light that was absorbed by a sample directly. We infer on the fraction of absorbed light by measuring light that was not absorbed.

1.3 THE SPECTROSCOPIC EXPERIMENT AND THE SPECTROMETER

Now that we have introduced the key elements of a spectroscopic experiment, as well as the reflectance and transmittance of a sample, it is possible to go back to the spectroscopic experiment and to examine it in more detail. Notice that we really do not know the absolute source intensity $I_0(\nu)$ that we used to calculate intensity. And also, we do not know the absolute transmitted intensity either. What we measure on the detector is the intensity of light emitted by the source $I_0(\nu)$ modified by the transmittance of that light through the spectrometer optics $\Pi(\nu)$. The detector converts light intensity into an electrical signal and that conversion embeds the spectral sensitivity $D(\nu)$ of the detector into the output signal since the detector may respond differently to different frequencies of light. The electrical signal produced by the detector may be further amplified before it is converted into a number by digitizing electronics. If the detector signal contains components modulated at different frequencies (such as with Fourier transform infrared spectrometers), the amplification $\Omega(\nu)$ may be different for different frequencies. Thus, the recorded detector signal (spectrum)

$$S_0(\nu) = I_0(\nu)\Pi(\nu)D(\nu)\Omega(\nu)$$

may exhibit little resemblance to the source intensity spectrum $I_0(\nu)$. On the other hand, the detector signal recorded with the sample in the beam is, by the same token, given by

$$S(\nu) = I_0(\nu)\Pi(\nu)D(\nu)\Omega(\nu)T(\nu),$$

where the effect of the sample is described by the sample transmittance $T(\nu)$. Note that the only difference between the spectrum recorded without the sample in the beam $S_0(\nu)$ and with the sample in the beam $S(\nu)$ is the transmittance factor $T(\nu)$. That means that, by ratioing the two spectra, all the other unknown quantities cancel out and what is obtained is the pure transmittance of the sample. Thus, to measure sample transmittance in a spectrometer, we do not need to know the details of detector response, the specifics of amplifier, the spectral intensity of light emitted by the source, the optical characteristics of the spectrometer optics, and so on. The only thing we need to know is that all these characteristics of the various spectrometer components are stable and reproducible so that when we take the ratio, they indeed all cancel out.

Even if all the components posses the needed stability and their characteristics indeed do cancel out, there are additional fine points that have to be accounted for in order to obtain the correct measurements. So far, it has been assumed that the insertion of the sample into a spectrometer beam did not have any effect on optical imaging of the spectrometer's optical components. Generally, if a sample is a very thin freestanding film, the optical consequences of its insertion into the beam are negligible. However, if the sample is thick, the optical effect of the sample insertion is a slight shift of the focal point of the beam and hence defocusing of the beam on the detector. This shift introduces a change in light intensity on the detector, an effect that is not present in the reference spectrum $S_0(\nu)$. Therefore, the measured transmittance is no longer the pure transmittance of the sample; it is modified by the spectrometer's beam defocusing caused by the presence of the sample in the beam of the spectrometer.

Almost everything said up to now about the measurement of transmittance applies to the measurement of reflectance. The reflectance measurement is even trickier, however, because the reflectance measurement involves change in the beam direction. The experimental design has to provide for a way to conduct a reflectance measurement. One way would be to move the detector together with the detector optics around and to position them to catch the reflected radiation. If attention is paid to the proper alignment of the detector optics, moving of the detector into the reflected light could be a viable strategy for reflection spectroscopy. However, most spectrometers do not provide for this functionality. One can see why. Such a setup would introduce a very hard to characterize alignment factor. Even the tiniest difference in alignment on the detector between the reference and the sample beam would affect light intensity on the detector and would reduce the accuracy of the reflectance measurement. So, in most cases, the reflectance measurement is done in a fixed optical configuration first with a reference, then with the sample (Fig. 1.2). In a standard setup, two

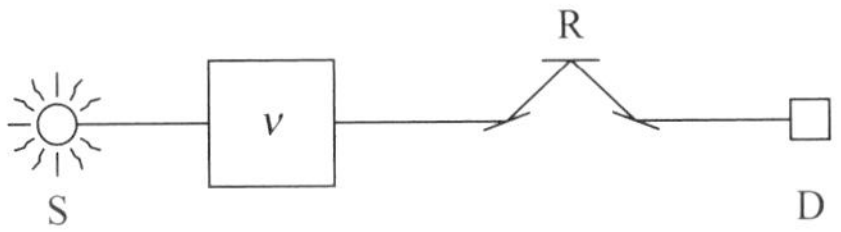

Figure 1.2 Reflectance measurement in a spectrometer.

mirrors are used to reflect the incident light to the sample and the reflected light back into the beam path to detector.

This setup removes possible misalignment between the reference and sample beams. Note that even if there is some misalignment in the reflectance setup, this does not become a problem as long as it is identical for the reference and sample beams. The ratio of the sample and reference spectra now yields the reflectance of the sample divided by the reflectance of the reference reflector, and a possible misalignment factor cancels out.

In many cases, a good metal mirror is used for reference and the result is taken as an adequate approximation of the sample reflectance. However, it must be emphasized that the reflectance measurement, unlike transmittance measurement, involves an additional unknown—the reflectance of the reference used.

If a high precision reflectance measurement is needed, then clearly, the above procedure would not do unless the reflectance of the reference is known. Those reflectance measurements where the precise reflectance of the sample must be measured are referred to as absolute reflectance measurements, and special techniques must be employed to perform such measurements.

Let us consider the process of measuring a spectrum in a bit more detail. First, note that two spectra need to be collected to obtain transmittance or reflectance of the sample. First, one collects the reference spectrum then inserts the sample and collects the sample spectrum. But notice how this is measured. What we consider to be the signal is really the absence of light at the detector. However, what the detector sees as a signal is the presence of light on the sensor. The detector produces an electrical signal proportional to the power of light on the detector element. In addition to the signal, the detector and the detector electronics produce noise. Noise is an inevitable consequence of the thermal fluctuations of electrical charges within the various electrical components.

A good electrical design can minimize the noise level produced by the detector and its associated electronics, but there is a minimum level of noise that just cannot be eliminated. For instance, a resistor is the source of the so-called Johnson noise—thermally induced voltage fluctuations on the terminals of the resistor. These fluctuations add to the signal produced by the detector. The quality of the measurement depends on the relative strengths of the signal component due to light intensity and the signal component due to noise. The ratio of signal to noise (S/N) is used as one measure of quality of the experimental results.

Since the noise is superimposed onto the signal at the detector level, it is impossible to separate them out later on. If an amplifier is used to amplify the detector signal, it amplifies both components of the signal, hence leaving their ratio unchanged.

1.4 PROPAGATION OF LIGHT THROUGH A MEDIUM

Although the following material is covered in many elementary texts on optics or electromagnetic theory, the need for a consistent notation and nomenclature justifies a brief reiteration of the subject. The knowledge of electromagnetic phenomena is condensed into four Maxwell equations. The Maxwell equations in a homogeneous medium without sources and characterized by the dielectric constant ε are

$$\begin{aligned} \nabla \cdot \mathbf{E} &= 0 \\ \nabla \cdot \mathbf{B} &= 0 \\ \nabla \times \mathbf{E} + \frac{1}{c}\frac{\partial \mathbf{B}}{\partial t} &= 0 \\ \nabla \times \mathbf{B} - \frac{\varepsilon}{c}\frac{\partial \mathbf{E}}{\partial t} &= 0, \end{aligned} \tag{1.3}$$

where **E** and **B** are the electric and magnetic field vectors. The equations (Eq. 1.3) combine into the same wave equation for either of the fields. For the electric field, we have

$$\left(\frac{\partial^2}{\partial \mathbf{x}^2} - \frac{\varepsilon}{c^2}\frac{\partial^2}{\partial t^2} \right) \mathbf{E}(\mathbf{x}, t) = 0, \tag{1.4}$$

where $\mathbf{E}(\boldsymbol{x}, t)$ is the amplitude of the electric field vector at the position $\boldsymbol{x}$ and time t, and c is the speed of light in vacuum. The solution to Equation 1.4 is a plane wave:

$$\mathbf{E}(\mathbf{x}, t) = \mathbf{E}_0 e^{2\pi i n \mathbf{k}\mathbf{x} - i\omega t}, \tag{1.5}$$

where $n = \sqrt{\varepsilon}$ is the refractive index of the medium; the wave vector **k** points in the direction of wave propagation and has a magnitude (also known as wave number) k given by

$$2\pi n k = n\frac{\omega}{c} = \frac{2\pi n}{\lambda}, \tag{1.6}$$

where ω is the angular frequency and λ is the wavelength of light in vacuum. Note that Equation 1.6 follows from the substitution of Equation 1.5 into Equation 1.4. In other words, for Equation 1.5 to be a solution of Equation 1.4, the wave number and the frequency of the plane wave (Eq. 1.5) have to be connected as indicated in Equation 1.6.

The same equation, and hence the same solution, holds for the magnetic field $\mathbf{B}(\boldsymbol{x}, t)$. The two plane waves are not independent; they propagate together and oscillate in phase in every point in space.

By inserting the solutions for **E** and **B** into the first two equations in Equation 1.3, one finds that both **E** and **B** are perpendicular to $\boldsymbol{k}$. By inserting the solutions (Eq. 1.5) for **E** and **B** into the third and fourth equation in Equation 1.3, we find that **E** and **B** are also perpendicular to each other.

The magnetic field can be expressed in terms of the electric field with the help of the third Maxwell equation in Equation 1.3 as

$$\boldsymbol{B} = \frac{2\pi cn}{\omega} \boldsymbol{k} \times \boldsymbol{E}. \tag{1.7}$$

Thus, light propagates through a medium characterized by the refractive index n essentially in the same way as through vacuum, except that the speed of propagation is reduced to c/n. As a consequence, the wavelength is shortened to λ/n (hence the wave number lengthened to nk).

Note that any direction of propagation is allowed by Equation 1.5. The proper direction(s) of propagation is (are) determined by the circumstances of a particular situation.

If the medium absorbs, light gets weaker as it penetrates into the medium. This is very easy to incorporate into the formalism by simply allowing the refractive index to become a complex number:

$$n_c = n + i\kappa. \tag{1.8}$$

The real and imaginary parts of a complex refractive index are generally referred to as the optical constants of the material. Using Equation 1.8 in Equation 1.5 yields

$$E(x,t) = E_0 e^{-2\pi\kappa kx} e^{2\pi nkx - i\omega t}. \tag{1.9}$$

The first exponential factor in Equation 1.9 describes the exponential decay that accounts for light absorption. The second exponential factor is the usual oscillatory factor responsible for the description of light propagation in the medium. The light intensity is proportional to the absolute value squared of the electric field, so the light intensity decays according to

$$I(x) = I(0)e^{-4\pi k\kappa x}, \tag{1.10}$$

which is known as Lambert's law and which explicitly describes the absorption of light as it propagates through an absorbing medium.

1.5 TRANSMITTANCE AND ABSORBANCE

The intensity of light exiting the material is reduced exponentially with the increasing thickness of the sample. As we have seen earlier, the factor giving the fraction of incident light that transmits through the sample of thickness d is called the transmittance of the sample:

$$T(d) = e^{-4\pi k\kappa d}. \tag{1.11}$$

The transmittance depends on the sample thickness d, on the wave number of light $k = 1/\lambda$ and on the absorption index κ.

The wave number of light k is by definition just the inverse of the wavelength. It physically represents the number of wavelengths of light that fit in a 1-cm length. So, in a sense, the wave number is the spatial frequency of light since it tells us how many wavelengths of light fit into a unit of length just as (temporal) frequency ν tells us how many periods of oscillations of light fit into a unit of time.

The absorption index κ is a property characteristic of the material through which light propagates. Generally, the absorption index is a function of the frequency of light.

As we saw in Equation 1.11, all the variables regulating the transmission of light through a sample are in the exponent. It is thus convenient to take the logarithm of the transmittance. For historical reasons, the logarithm is not a natural logarithm but a decadic logarithm. Taking the decadic logarithm of both sides of Equation 1.11 gives the expression

$$A = -\log(T) = 4\pi k\kappa d \cdot \log(e). \tag{1.12}$$

The negative logarithm of transmittance is known as the absorbance of the sample. As we can see, the absorbance is a linear function of the sample thickness. Sample thickness is an incidental property of a particular sample, not a characteristic property of the material of which this sample is composed. So, the absorbance of a sample of thickness d is often divided by the sample thickness to arrive at the so-called absorption coefficient of the sample $\alpha(k)$, which is a characteristic property of the material.

In most cases, it is a good assumption that the absorption coefficient of a sample is proportional to the concentration of the sample. This assumption, universally known as Beer's law, is the foundation for the analytical use of spectroscopy. As we will see later, Beer's law is not exact in the sense in which Equation 1.12 is. It is rather an approximate relation deriving from the fact that over a limited range, every function can be approximated by a linear function. Despite its approximate nature, Beer's law holds surprisingly well in a vast number of practical cases.

The concept of absorbance is so deeply ingrained into the spectroscopic culture that it is often seen as the quantity directly measured in a spectroscopic experiment. However, absorbance is a derived quantity. In an experiment, two single-beam spectra are measured, one a reference and the other the sample spectrum. The two spectra are divided to yield the transmittance. Although the transmittance is thus not a directly measured quantity but is rather a consequence of postmeasurement data manipulation, transmittance is usually seen as the direct result of a measurement. However, absorbance requires a significant mathematical manipulation of the experimental data. Thus, absorbance is a transform of the original experimental data. This transform of the transmittance is performed because it leads to the quantity that is a linear function of both sample thickness and, within the jurisdiction of Beer's law, the sample concentration. Taking a logarithm is quite a nonlinear transform of the original data. In the spectroscopic measurement, it is the absorbance that is seen as the "signal" measured in the experiment. This is so because, according to Beer's law, the absorbance is proportional to the concentration of the material, and it is the determination of concentration that is the ultimate goal of spectroscopic measurements. So let us see how this peculiar definition of signal affects the signal-to-noise ratio of a spectroscopic measurement.

1.6 S/N IN A SPECTROSCOPIC MEASUREMENT

When a single-beam spectrum $S(k)$ is measured, both the "true" signal $S_0(k)$ and the noise $\delta(k)$ are contained in the measured value; that is,

$$S(k) = S_0(k) + \delta(k). \tag{1.13}$$

Let us, for the time being, ignore the precise details of the measurement and focus on the propagation of noise as we advance from single beams toward absorbance. First, we assume that noise is random and that it can take both negative and positive values with equal probability. Thus, the average value of noise readings is zero:

$$\langle \delta(k) \rangle = 0. \tag{1.14}$$

The noise is randomly varying as the detector signal is measured and converted into a number (digitized). As a measurement is made, each value associated with a particular wave number gets its own random value of noise. So it looks as if the noise is a function of wave number. Noise, however, is completely random. We associate the noise value with wave number in Equation 1.13 to indicate that each measured value of signal gets its own value of noise. Noise values, if truly random, are not correlated; that is, the noise values for two consecutive wave numbers are not connected in any way, and if the measurement is repeated under identical conditions, the noise values at the same wave number would not be correlated. Those are our assumptions about the noise. These assumptions are not necessarily fulfilled in every experimental situation.

We, of course cannot measure noise in an experiment because if we could, we would simply subtract it out from the measured values and get a perfect noiseless spectrum. However, we can do something close to measuring the noise and at least learn about how it affects our experiment. We can repeat the measurement for $S_0(k)$ a large number of times, say, N, under the same conditions, find the average value at each wave number, and then go back and subtract this average value from each of the experimental values. The result of this subtraction is a good approximation for the actual experimental noise in a particular spectrum.

Now we can analyze the noise itself. We can focus on a specific wave number k and list the sequence of noise values obtained in the multiple measurements:

$$\delta_1(k), \delta_2(k), \ldots, \delta_N(k). \tag{1.15}$$

To characterize the strength of the noise, we cannot take an average value of the sequence (Eq. 1.15) since this is zero according to Equation 1.14. A good way to measure the strength of noise is to take the average of the squares of the noise values in Equation 1.15:

$$\langle \delta^2(k) \rangle = \sigma^2(k). \tag{1.16}$$

The value $\sigma(k)$ measures the strength of the noise. As we have already said, we do not really expect the noise to depend on the wave number, so we could drop the dependence of σ on k in Equation 1.16.

We also see that the signal to noise of a single-beam measurement is just

$$S/N = S(k)/\sigma, \tag{1.17}$$

that is, the S/N follows the shape of the single-beam spectrum.

We can now calculate the noise in a transmittance measurement. We collect the background single beam $S(k)$ and the sample single beam $T(k)S(k)$. To get the transmittance of the sample $T(k)$, we divide the sample single beam with the background single beam. But, as we have seen, in collecting the single-beam spectra, the noise is added to the true experimental values, so

$$\begin{aligned} S_B(k) &= S_B 0(k) + \delta(k) \\ S_T(k) &= T_0(k) S_{B0}(k) + \delta'(k), \end{aligned} \tag{1.18}$$

where the prime on the second noise value indicates that, because the second measurement collects a different noise sequence, the noise values are not the same, although they have the same strength (Eq. 1.16).

The subscript zero on the sample transmittance indicates the true transmittance. The measured transmittance differs from the true transmittance because of the presence of noise. The ratio, after some manipulation, yields

$$T(k) = T_0(k) + \frac{\sigma}{S_0(k)} \sqrt{1 + T_0^2(k)}. \tag{1.19}$$

Here, we kept only the term linear in noise strength. Thus, the noise at the level of transmittance picks up a weak dependence on the transmittance measured. Since the square root term in Equation 1.19 varies between 1 and 1.41 as transmittance varies from 0 to 1, this dependence of noise on transmittance is generally ignored.

Thus, we see that the noise in the single-beam measurement of strength σ translates into the noise in transmittance measurement as

$$\delta T(k) = \frac{\sigma}{S_0(k)} \sqrt{1 + T_0^2(k)}.$$

To see how noise propagates through absorbance transform, we start by rewriting Equation 1.12 as

$$T = 10^{-A}. \tag{1.20}$$

Now, we can write

$$\delta A = -\frac{1}{\ln 10} 10^{A} \delta T. \tag{1.21}$$

Thus,

$$\frac{A}{\delta A} = 2.303 \frac{S_0}{\sigma} \frac{A \cdot 10^{-A}}{\sqrt{1 + 10^{-2A}}}, \tag{1.22}$$

where 2.303 = ln(10). Also, we dropped the sign from Equation 1.21 since we are interested in magnitudes only. Equation 1.22 shows how the S/N of an absorbance measurement depends on the absorbance measured.

We see that, for a weak absorbance, S/N increases linearly with absorbance. For a strong absorbance, the exponential term in Equation 1.22 takes over and the S/N quickly decays to zero. For the absorbance value of $A = 0.4$, S/N is maximized. This differs from what was found for the single-beam S/N ratio (Eq. 1.17) where the stronger the single-beam signal, the higher the S/N of the measurement.

This characteristic of the S/N of an absorbance measurement has to be kept in mind when designing an experiment. The absorbance level measured in an experiment can often be selected. In ATR and transmission measurements, a number of experimental parameters can be tuned to set the level of absorbance measured in the experiment. Whenever the level of absorbance measured in an experiment can be influenced by the selection of experimental parameters, Equation 1.22 provides the guidance for optimizing the S/N.

2 Harmonic Oscillator Model for Optical Constants

2.1 HARMONIC OSCILLATOR MODEL FOR POLARIZABILITY

As we have seen, the propagation of light through a medium is controlled by the refractive index of the medium. The medium can be transparent to light or it can absorb light. We have seen that by allowing the medium's refractive index to be complex, that is, to have both real and imaginary components, the model can easily accommodate the absorption of light by the medium. This may have looked like a clever trick, but we will show now that there is an unexpected depth behind the trick—a model for the refractive index of a medium that, in simple and clear terms, encompasses virtually all the observed phenomenology of optical spectroscopy.

First, the medium through which light propagates is really not continuous—it consists of molecules. Molecules, in turn, are built from atoms held together by chemical bonds. Chemical bonds between atoms act to hold atoms together. These bonds are strongly repulsive if the atoms are very close but turn attractive as the atoms are farther apart. Thus, there is a point at which the force between two atoms vanishes, and if the atoms are from that point pushed closer together, a repulsive force pushes them apart; if they are pulled apart, the attractive force pulls them toward each other.

So, the atoms behave as if they are connected by springs as shown in Figure 2.1. The potential energy between two atoms as a function of distance looks as shown in Figure 2.2.

Point *a* marks the minimum of potential energy. Every function can be approximated by a parabola near its minimum. However, this parabolic approximation is adequate only for small deviations around the minimum. The minimum point is also the equilibrium point, that is, the

---

Internal Reflection and ATR Spectroscopy, First Edition. Milan Milosevic.

Figure 2.1 Two masses connected with a spring.

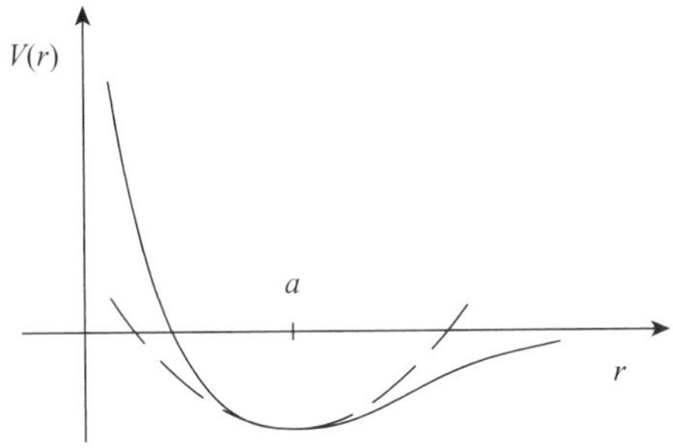

Figure 2.2 Potential energy of two atoms as a function of distance. The dashed line shows a parabola osculatory to the potential in the minimum.

point in which the force between the two atoms is zero. But the atoms are not stationary; they possess thermal energy and thus oscillate around the equilibrium point. For small oscillations, the shape of the potential energy around the minimum is indistinguishable from a parabola. Thus, for small oscillations around the minimum, the chemical bond between two atoms in a molecule could be modeled as if it was a tiny spring. Note that this conclusion is true regardless of what the exact shape of the potential energy looks like. It simply follows from the requirement that the potential energy has a minimum point. The oscillations of atoms connected by such bonds are thus, in first approximation, harmonic.

A molecule has a number of bonds binding the constituent atoms together. Bonds between different atoms generally have different strengths, and the masses of atoms participating in the bond are different, so the frequencies of oscillations of the different atoms in a molecule around their equilibrium positions are different for different bonds and are characteristic of the type of the bond. A good way to picture a molecule is to imagine atoms thermally excited into a symphony of oscillations around their respective equilibrium points, each bond oscillating with its own characteristic frequency. An atom in a molecule participates in more than one bond, and the actual oscillations are a complex superposition of the various interacting oscillations. Some atoms are strongly bonded together, some weakly. The effects of oscillations of any single atom are, through these bonds, transferred to any other atom in the molecule, and the oscillations of every atom become complex. If the bonds between the atoms in a molecule were all purely harmonic, the description of a complex system of interacting atoms could be transformed into a description of a collection of non-interacting harmonic oscillators called normal modes.

Each normal mode is a complex superposition of individual atomic vibrations. The miracle of the normal mode transformation is that each such superposition oscillates by its own frequency unaffected by how strongly or weakly the other normal modes are oscillating. This normal mode transformation is possible only for pure harmonic bonds between atoms. Real bonds are not harmonic and a result is that the normal modes are not completely independent but weakly interact, causing weak molecular vibrations at frequencies that are multiples and combinations of the fundamental frequencies associated with normal modes.

Let us see how this normal mode transformation comes about. Imagine a molecule consisting of a number of atoms. Such a configuration of atoms is bonded together by chemical bonds between the atoms. Each atom is located in the equilibrium point of the various forces acting on it. Because of thermal motions, an atom is not stationary in its equilibrium point but oscillates around it.

The potential energy of a molecule is a function of the positions of the constituent atoms:

$$U = U(q_1, q_2, \ldots, q_b) \tag{2.1}$$

where $q_1, q_2, \ldots, q_b$ are the coordinates of the atoms and b is the number of degrees of freedom of the molecule. If we denote the coordinates of the equilibrium position as $q_1^0, q_2^0, \ldots, q_b^0$ and the deviation of the ith coordinate from its equilibrium point as $x_i = q_i - q_i^0$, we can express the potential energy (Eq. 2.1) as a Taylor series in terms of small deviations x_i. The result is

$$U = U_0 + \frac{1}{2}\sum_{i,j} \frac{\partial^2 U}{\partial x_i \partial x_j} x_i x_j + \cdots. \tag{2.2}$$

Note that the linear terms in the Taylor expansion vanish since the first derivatives of the potential energy are zero at the minimum. Higher order terms are small, but nonvanishing, and, as we said above, they are responsible for the weak vibrations that are multiples and combinations of the fundamental vibrations. The energy of the molecule is then simply a sum of the kinetic and potential energies of the individual vibrations.

Ignoring the higher order terms in Equation 2.2, the energy is simply

$$E = \frac{1}{2}\sum_i M_i \left(\frac{dx_i}{dt}\right)^2 + \frac{1}{2}\sum_{i,j} \frac{\partial^2 U}{\partial x_i \partial x_j} x_i x_j, \tag{2.3}$$

where M_i is the mass of the mode associated with the ith coordinate. Clearly, the equations of motions derived from Equation 2.3 are coupled. But, since the term $\pi_{ij} = \partial^2 U/\partial x_i \partial x_j$ is symmetric in indices i and j, that is, $\pi_{ij} = \pi_{ji}$, a unitary transformation of the coordinates V,

$$X_i = \sum_\alpha V_{i\alpha} x_\alpha, \tag{2.4}$$

exists, such that

$$E = \frac{1}{2}\sum_i M_i \left(\frac{dX_i}{dt}\right)^2 + \frac{1}{2}\sum_i \frac{\partial^2 U}{\partial X_i^2} X_i^2. \tag{2.5}$$

Now the vibrational energy of the molecule is just a sum of the independent vibrational energies of these collective modes of vibrations. By a comparison with the expression for the energy of oscillations of a simple harmonic oscillator, it follows that the second derivative terms in Equation 2.5 are just the squares of frequencies of vibrations of the normal modes.

What happens if a molecule is placed in an electric field? Obviously, the electrons of the bound atoms are pushed one way, while the nuclei are pushed the other way. In general, such a distortion of electric charge causes a dipole moment. The electric charges in the molecule may be distributed in such a way that the molecule has a dipole moment even in the absence of an external electric field. However, for present purposes, we are interested in the molecular dipole moment that is induced by an external electric field. In the absence of an external field, the positive and negative charges are centered on the same point, and thus there is no dipole moment. The state of zero dipole moment is the equilibrium state and, as we have already seen in discussing chemical bonds, a small displacement from the equilibrium will necessarily cause a harmonic restoring force. As an external field segregates the charges inducing a dipole moment, the harmonic force acts on these charges providing the restoring force that pulls them back together.

The equation for a single normal mode in the presence of an external electric field **E** is

$$\mathbf{E}q = m\left(\frac{d^2\mathbf{x}}{dt^2} + \gamma\frac{d\mathbf{x}}{dt} + \omega_0^2\mathbf{X}\right), \tag{2.6}$$

where q is the charge associated with the normal mode, ω_0 is the natural frequency of the harmonic oscillations of the normal mode, m is the mass of the normal mode, and γ is the damping constant.

The only new element introduced in Equation 2.6 is the second term associated with the dissipation of the vibrational energy of the mode. It is intuitively clear that if there is an excess of energy in a particular normal mode, this excess is going to dissipate to other normal modes of the same molecule or to other molecules. The mechanism of this dissipation may be complex, that is, taking place through higher order terms in the molecular potential, through interactions with other molecules, and so on. From thermodynamics it is known that, in a thermal equilibrium, thermal energy is equally distributed among all the degrees of freedom. So thermodynamics requires that there exists a mechanism that equalizes energy between different degrees of freedom. The true form of the dissipative term in Equation 2.6 is certainly not the one shown, but the form of dissipation used in Equation 2.6 should be able to mimic the proper dissipation closely enough for the present purpose. It essentially states that there is a frictional force proportional to the speed of the mass m, which resists motion of the oscillator, thus damping the oscillations.

If the external electric field $\mathbf{E} = \mathbf{E}_0 e^{i\omega t}$ is the electric field of an electromagnetic wave oscillating with frequency ω, then the solution for $\mathbf{X}$ is expected to oscillate with the same frequency, that is,

$$\mathbf{X} = \mathbf{X}_0 e^{i\omega t}, \tag{2.7}$$

where $\mathbf{X}_0$ is the amplitude of the oscillations. Note that above, we ignored the e^{ikx} factor in the expression for the plane wave. We could do this since the oscillations of atoms around their equilibrium positions in a molecule are extremely small compared to the wavelength of light. Therefore, the electric field of the electromagnetic wave is essentially the same in the vicinity of the equilibrium point as it is in the point itself.

By inserting Equation 2.7 into Equation 2.6, we can find the amplitude of oscillations:

$$\mathbf{X}_0 \frac{q/m}{\omega_0^2 - \omega^2 + i\gamma\omega}\mathbf{E}. \tag{2.8}$$

As we can see from Equation 2.8, the amplitude of oscillations of the normal mode oscillator is a function of the frequency of the electromagnetic field. The oscillations represent movements of the electric charge q around the equilibrium point which generates an oscillating dipole moment $\mathbf{p} = q\mathbf{X}$. The polarizability $\alpha(\omega)$ is defined as the ratio of the induced dipole moment and the applied electric field, so we have

$$\alpha(\omega) = \frac{q^2/m}{\omega_0^2 - \omega^2 + i\gamma\omega}. \tag{2.9}$$

The amplitude has a resonant form that is peaked at the natural frequency of oscillations of the normal mode. This shows that the maximum polarizability of a molecule occurs near resonant frequencies of its normal modes. Since normal modes are mutually independent, the polarizability of a molecule is the sum over the normal modes:

$$\alpha(\omega) = \alpha_0 + \frac{e^2}{m_0} \sum_{j=1}^{b} \frac{f_j/\mu_j}{\omega_j^2 - \omega^2 + i\gamma_j\omega}, \tag{2.10}$$

where μ_j is the dimensionless factor relating the mass of proton m_0 and the mass associated with jth normal mode m_j, e is the charge of electron, and f_j is the so-called oscillator strength. The constant α_0 is the polarizability that is due to electrons and is not really a constant but a function of frequency. However, at frequencies associated with bond vibrations, this term is essentially constant. This is because electrons are much lighter than atoms and the resonant frequencies for electronic vibrations are much higher than those for molecular vibrations. Thus, the electronic contribution to polarizability at molecular vibration frequencies can be approximated by a constant.

Another point is that our equations contain the imaginary constant i. Of course, using complex numbers is just a mathematical convenience. It makes expressions much more compact. However, the expressions for measurable quantities or observables must always be real.

2.2 CLAUSIUS–MOSSOTTI EQUATION

A medium consisting of identical molecules, each having the polarizability given by Equation 2.10 and having a density (number of molecules per unit volume) N will possess a dielectric constant, $\varepsilon(\omega)$. According to the Clausius–Mossotti relation, the dielectric constant of a material is related to the polarizability of an individual molecule and density of molecules in the material as

$$\varepsilon(\omega) = 1 + \frac{4\pi N\alpha(\omega)}{1 - \frac{4\pi N\alpha(\omega)}{3}}. \tag{2.11}$$

A molecule in a medium responds to the electric field that acts on it. This electric field is not just the applied external electric field but also includes the electric fields of the induced dipoles of the surrounding molecules. Since the induced dipoles of the surrounding molecules are also responding to the total electric field acting on them, there is a complex feedback between the external electric field and the fields due to induced dipole moments of the surrounding molecules. The above expression (Eq. 2.11) sums up this feedback. What it does not take into account are the changes to the frequencies, the oscillator strengths, and the effective masses of normal modes that are caused by the intermolecular interactions.

2.3 REFRACTIVE INDEX

The refractive index of a medium is the square root of the dielectric constant, so the expression (Eq. 2.11) gives us a model for the refractive index. Note that the expression for the dielectric constant gives complex number values for the dielectric constant, and the model thus yields complex number values for the refractive index. As we have seen before, the absorption of light by a medium is controlled by the imaginary part of the refractive index. We modeled the normal mode oscillations in Equation 2.6 by adding the oscillation damping term to a standard harmonic oscillator. This damping force is proportional to the velocity of the oscillating mass not unlike the damping that a fluid, such as water, would cause to a ball attached to a spring and immersed in it. This model attributes the absorption of the oscillator energy to friction. This is obviously not what is happening in the case of a molecular oscillator, but it could nevertheless be used to model the complex processes that dissipate energy from molecular oscillations.

Before we examine the implications of the harmonic oscillator model for molecular polarizability, let us first reexpress Equations 2.10 and 2.11 in terms of the wave number k instead of the angular frequency of light ω. By definition, $k = 1/\lambda$, and by using Equation 1.6, we can substitute

$$\omega = 2\pi ck \tag{2.12}$$

and

$$\gamma = 2\pi c\Gamma \tag{2.13}$$

into Equation 2.10 to get the expression for molecular polarizability in terms of wave numbers. This we do for convenience since infrared (IR) spectra are almost exclusively displayed with wave numbers on the abscissa axis. Thus, we get

$$\alpha(k) = \alpha_0 + \frac{e^2}{(2\pi c)^2 m_0} \sum_{j=1}^{b} \frac{f_j / \mu_j}{k_j^2 - k^2 + i\Gamma_j k}. \tag{2.14}$$

The expression under the summation sign in Equation 2.14 is complex and could be separated into real and imaginary parts. The imaginary part is, as always, related with absorption. The imaginary part of a typical term in the sum is

$$\frac{\Gamma_j k}{(k_j^2 - k^2)^2 + \Gamma_j^2 k^2}. \tag{2.15}$$

The meaning of the parameter Γ, which is by definition (Eq. 2.13) related to the damping constant of the harmonic oscillator model γ, becomes clearer if we observe that the expression has the shape of a peak centered on $k = k_j$ and that it quickly falls to zero as k moves away from k_j. The peak maximum is $1/\Gamma_j k_j$, so everything else being the same, the maximum is smaller for peaks at higher wave numbers and with greater value of Γ. Next, we see that Γ controls the width of the peak. If $k = k_j \pm \Gamma/2$, the value of the expression falls very close to ½ of its value at the peak. That means that the width of the peak at half height is Γ. Thus, the constant responsible for the damping of the oscillations controls both the height and the width of the absorption peak. Another term in Equation 2.14 is the ratio f_j/μ_j, which can be replaced by a single constant, F_j. This constant measures the strength of a particular mode.

The value of the harmonic oscillator model to spectroscopy hinges on its ability to model the refractive index of actual materials. The potential value of the harmonic oscillator model for modeling the refractive index of a medium is that every peak observed in the spectrum can be described by only three parameters: strength F, width Γ, and the peak position k_0. This seems very promising since the peak position and width can usually be read off from the spectrum and only the strength has to be adjusted to reproduce the height of the observed peak.

Let us see how the model works, that is, how realistic the optical constants that the model produces are when compared to optical constants of real materials. To proceed, let us consolidate all the constants outside the summation symbol in Equation 2.14 and combine them with density N that multiplies molecular polarizability in Equation 2.11 into a single constant, c_0. The exact value of this constant is not that important since for modeling purposes, it is the product of c_0 and the strengths F_j that matter. We could pick a value, say, $c_0 = 1.2 \times 10^5$. This more or less sets the order of magnitude of the expression. Now we still have to pick out the value for the electronic polarizability. A way to pick that is to assume that the vibrational term in Equation 2.14 vanishes and to find the value of α_0 that would lead to the value of the refractive index of 1.5, which is typical for most materials. Simple algebra yields $\alpha_0 = 0.8$. Now let us pick the parameters for a peak. Since we are interested in the mid-IR spectral region (400–4000 cm^{-1}), we can pick the peak position as $k_0 = 1000\,\text{cm}^{-1}$, the peak width $\Gamma = 10\,\text{cm}^{-1}$, and the strength $F = 0.1$. We insert these values into Equation 2.14, then insert that into Equation 2.11 to get the dielectric constant, and then we take the square root of the dielectric constant to get optical constants. The above choices for the numerical values of the model parameters lead to the optical constants shown in Figure 2.3.

The wave number axis in Figure 2.3 is shown in reversed order as it is customary for displaying spectra in the mid-IR spectral region. The

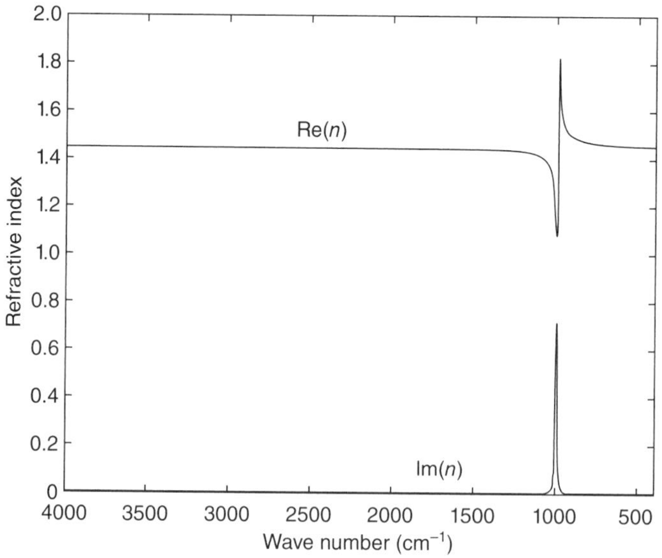

Figure 2.3 Modeled optical constants.

optical constants of materials are not directly measurable in a spectroscopic experiment, so we will have to utilize optical constants to model a directly measurable spectroscopic observable such as reflectance or transmittance to see how realistic the model is. For now, we notice that the model predicts the real part of the refractive index to sharply change in the vicinity of the resonant frequency and that the absorption index has a sharp peak at the resonant frequency k_0 while essentially zero elsewhere. The width of the peak is not quite Γ since the original shape (Eq. 2.15), also known as Lorentzian shape, gets modified in going through the above indicated steps. However, the optical constants shown in Figure 2.3 appear realistic. Therefore, we can conclude that the harmonic oscillator model yields realistic-looking optical constants and, therefore, it could be used to model optical constants of real substances. In the rest of this book, we will frequently generate optical constants using the above model by simply selecting a number of arbitrary peaks. We will generally not try to reproduce optical constants of any specific substance. However, once the optical constants are given, the calculations of spectroscopic observables like reflectance or transmittance will follow the exact laws of electromagnetism. Therefore, we will treat the harmonic oscillator model as a convenient way to generate realistic optical constants.

2.4 ABSORPTION INDEX AND CONCENTRATION

We have seen that the absorbance of a sample is defined as the negative decadic logarithm of measured transmittance. This definition was advanced in expectation that absorbance defined in this way would be a linear function of concentration. It is now possible to investigate this conjecture, also known as Beer's law, within the harmonic oscillator model. A look over the steps taken to arrive at optical constants clearly indicates that the relationship is not strictly linear. Specifically, the Clausius–Mossotti equation explicitly indicates that the relationship between density and dielectric constant is not linear. This nonlinearity is connected to the taking into account that polarized molecules influence other polarized molecules around them, thus accounting for a portion of intermolecular interactions. The next step, taking the square root of the dielectric constant to get the optical constants, introduces further nonlinearities. Thus, on the face of it, it seems highly unlikely that the relationship between the absorption index and the concentration would end up anywhere near linear. We can proceed with the above model and plot the absorption index for concentrations

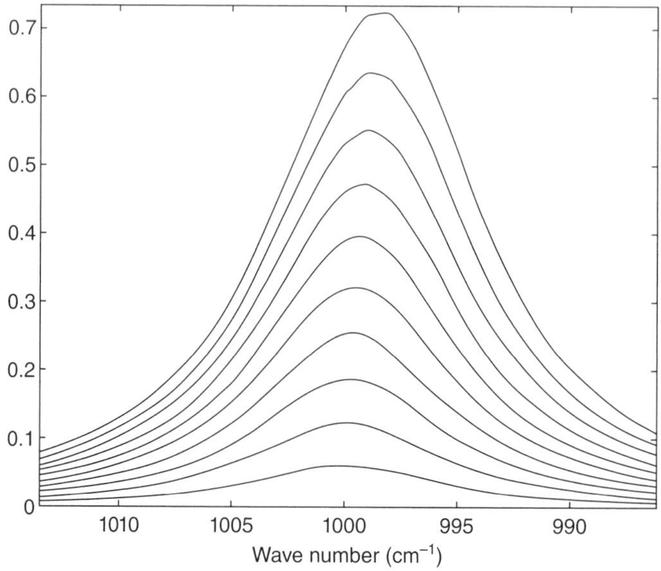

Figure 2.4 Dependence of the absorption index on concentration.

increasing in steps of 10% of the final concentration c_0. The result is shown in Figure 2.4. We see that, while the steps of increase of the absorption index are not identical, surprisingly, they are reasonably close to equidistant. A degree of nonlinearity is, however, apparent. Also, it is clear from Figure 2.4 that, with increasing concentration, the apparent peak position shifts to lower wave numbers. The shift is small but clearly noticeable.

Since, according to Equation 2.10, the dependence of absorbance on concentration derives from the dependence of the absorption index on concentration, the same nonlinearity observed with the absorption index is passed on to the absorbance. In addition, the expression (Eq. 2.9) from which Equation 2.10 was derived is a pure material property and does not fully reflect the behavior of absorbance derived from experimentally measured transmittance. In particular, it does not reflect reflectance losses that occur at the sample surfaces and that are incorporated into measured transmittance. Taking these reflections into account further impacts the nature of the dependence of measured absorbance on concentration. The conclusion, however, is that despite the unpromising look of the various expressions leading from molecular polarizability to absorption index, the dependence of absorbance on concentration is not that far from being linear, as assumed by Beer's law, at least for the substances whose optical constants can be modeled by the harmonic oscillator model.

3 Propagation of Electromagnetic Energy

3.1 POYNTING VECTOR AND FLOW OF ELECTROMAGNETIC ENERGY

It can be shown by formal manipulation of Maxwell equations for the region free of sources and currents that the following equation is true:

$$\frac{\partial}{\partial t}\left[\frac{1}{8\pi}\left(E^2+B^2\right)\right]+\nabla\left[\frac{c}{4\pi}(\mathbf{E}\times\mathbf{B})\right]=0. \tag{3.1}$$

The interpretation given to the expression (Eq. 3.1) is that it describes the energy conservation of the electromagnetic field. The term

$$u=\frac{1}{8\pi}\left(E^2+B^2\right) \tag{3.2}$$

expresses the energy density of the electromagnetic field in terms of field strengths. Notice that, for electromagnetic waves, the electric and magnetic fields have the same magnitudes, that is, $B^2=E^2$, so

$$u=\frac{1}{4\pi}E^2.$$

The vector quantity

$$\boldsymbol{P}=\frac{c}{4\pi}(\mathbf{E}\times\mathbf{B}) \tag{3.3}$$

is the so-called Poynting vector, which describes the flow of energy contained in the electromagnetic field. Since for electromagnetic waves

---

the electric and magnetic fields are mutually perpendicular and have the same magnitudes, the cross product of the fields in Equation 3.3 has a magnitude of E^2. The Poynting vector points in the direction of propagation of the electromagnetic wave and has the magnitude given by the product of the speed of light and energy density. Thus, Equation 3.1 can be rewritten as

$$\frac{\partial u}{\partial t} + \nabla \boldsymbol{P} = 0, \tag{3.4}$$

which simply states that the gain or loss of electromagnetic energy in a volume of space is due to the net flow of the electromagnetic energy in or out of that volume. Of course, that is exactly what one would expect based on the conservation of energy. There is nothing controversial here.

But there is something intriguing and a bit unnerving in the definition of the Poynting vector (Eq. 3.3) since the definition clearly associates the flow of electromagnetic energy with the mere presence of the two fields. Nothing has to oscillate; even static fields produce energy flow. A simple arrangement of static fields such as a magnetic field from a horseshoe magnet, an electric field from a capacitor with the two fields arranged perpendicular to each other, produces a mysterious flow of energy. It turns out, when carefully analyzed, that such arrangements of stationary fields could not be used as energy sources. No net energy flows in/out of any point in space. Energy flow just circles in closed loops. So what does such energy flow really mean and what is it that flows that remains mysterious? Most people ignore this, but some people are uneasy about it. It is not that people are uneasy with electromagnetic field possessing energy; it is that a static arrangement of fields produces a dynamic quantity—flow of electromagnetic energy, which is normally associated with the propagation of electromagnetic waves. It intuitively seems odd. While it is not of importance for our discussion, the unusual feature of the Poynting vector is mentioned in order to show that there are still a few murky topics in the electromagnetic theory.

Our concept of field ultimately derives from our concept of force. We can measure the electric field in a point by bringing a small unit charge to that point and measuring the force (both magnitude and direction) on the charge. In this way, we associate the electric field vector (which is, by definition, the force on unit charge) with that point in space. But how is space in that point different from space in any other point? What is the quality of space that is different when the electric field is present from when the field is not present? These are deep and unanswered questions. And if we cannot fully grasp the nature of the electromagnetic field, the derived quantities such as the Poynting vector can only be less intuitively transparent.

In the above, we focused on the Poynting vector as it relates to fields in empty space. If a medium is present, the expression for the Poynting vector becomes slightly more complicated in order to include the response of the medium to the fields present.

The importance of the Poynting vector to spectroscopy is in that it represents the energy flow contained in the electromagnetic radiation used in a spectroscopic experiment. Light emitted by a source, such as a very hot body, spreads in all directions from the surface of the source. The intensity of light emitted by a source is a function of the temperature of the emitting surface. If we imagine a sphere around the source, all the light from the source has to flow through the surface of the sphere. Every second, a certain amount of energy flows through the surface of this imaginary sphere. Energy emitted by a source per unit time is, by definition, the power of the source. By emitting energy, the source cools down and the light output of the source starts weakening. If we want to keep the light output of the source at a constant level, we need to feed the lost energy back to the source and we need to do it at the rate the source is emitting energy. When this is so, a certain constant light energy flows through the surface of the imaginary sphere. If the sphere is large compared to the source in its center, the flow of energy through every point on the sphere is the same. Power crossing per unit area of the surface of the sphere is the magnitude of the Poynting vector of the emitted radiation on the point on the sphere. The overall source output is then the surface area of the sphere multiplied by the magnitude of the Poynting vector. Since energy is conserved, this product is the same regardless of the radius of the sphere, implying that the magnitude of the Poynting vector of the emitted radiation falls off as a square of the radius of the sphere r:

$$P(r) = \frac{c}{r^2}. \tag{3.5}$$

The magnitude of the Poynting vector is more familiarly referred to as light intensity. If a portion of the radiation emitted by the source is collimated with a lens, the Poynting vector in a collimated beam is constant and no longer diminishes with distance from the source. If such a collimated beam is focused by another lens, the Poynting vector for the converging beam increases toward, and has a maximum value in, the focal point. The reason for bringing this up is that, in what follows, we will mostly use plane waves, which are of course associated with the propagation of collimated light beams. Strictly speaking, plane waves have infinite extension in space and time. Wave fronts of plane waves are planes perpendicular to the direction of propagation and parallel to each other. When we make a collimated beam by using a lens or a mirror, the beam size is limited by the size of the lens or mirror. Most often, the collimated beam is cylindrical in shape and the axis of the cylinder is what is called the optical axis or central ray of the beam. The wave fronts in the collimated beam are then equidistant parallel circles centered on and perpendicular to the central ray. However, we will still describe a collimated beam by using the concept of plane wave, so we will still refer to the optical axis of the beam, although the concept does not make sense for infinitely extended plane waves. We assume that since the wavelength of light is usually so much smaller than the diameter of a collimated beam, the plane waves are an excellent approximation for the description of such beams.

3.2 LINEAR MOMENTUM OF LIGHT

It is possible to define the stress tensor associated with electromagnetic fields in a point in space. This can be used to calculate the linear momentum associated with the flow of energy contained in light. The topic is outside the scope of this work, but it is nevertheless important to state that analysis leads to the conclusion that the Poynting vector describes the density of linear momentum carried by light divided by the square of the speed of light. The linear momentum of a particle is localized on the particle. Both the energy and the momentum of an electromagnetic wave are distributed in space just as is the electromagnetic field. We thus deal with energy density u and momentum density p_v. The energy and momentum are associated with a certain volume of space that travels with light and can be found by integrating the associated densities over the volume. Thus, not only does a traveling electromagnetic wave carry with it energy but, along with energy, it also carries linear momentum. To add to what we said before about nobody really knowing what makes a point in space different when an electric field is present from when it is not, and what is it that gives it energy, we now ask: What is giving it a momentum? Usually, momentum is associated with matter moving. What is interesting is that the relationship between energy and linear momentum for the electromagnetic wave resulting from these considerations is

$$u = p_v c. \tag{3.6}$$

This expression tells us that light possesses linear momentum and predicts that, for instance, as light reflects from a mirror, it imparts linear momentum to the mirror. Note that this is the equation that arises entirely out of the classical electromagnetic theory. Its form, however, anticipates both the relativistic connection between linear momentum and energy, $E = \sqrt{m^2c^4 + p^2c^2}$ (for massless particles such as a photon, $m=0$), as well as the quantum mechanical connection between frequency and energy, $E = h\nu$, and consequently, also wavelength and linear momentum ($p = hk = h/\lambda$), where h is the Planck constant.

3.3 LIGHT ABSORPTION IN ABSORBING MEDIA

As light propagates through an absorbing medium, its energy is absorbed by the medium and its intensity diminishes with the distance traveled. We have already seen that this is easily added to the existing formalism by allowing the refractive index of the medium to become a complex number. We have also seen how the harmonic oscillator (HO) model for refractive index naturally leads to complex refractive index near the resonant frequency of oscillations of a molecular bond. As we have seen, the HO model is a realistic description of molecular oscillations and the interaction of molecules with light. We also pointed out that the only artificial ingredient in the HO model is the description of the relaxation of molecular excitations. We used a simple model in which the excited molecular motion was damped by a friction-like interaction. Thus, in the HO model, the mechanism of light absorption is the work that incident electromagnetic field does against this force of friction. As a result, the electromagnetic energy is through this friction converted into heat. Although this friction model of light absorption is not realistic, the end result, that the energy of the absorbed light is converted into heat, is certainly correct. In a medium at ambient temperature T, all the degrees of freedom carry, on average, energy $k_B T_a/2$, where k_B is the Boltzmann constant and T_a is the absolute temperature of the medium. If light propagates through such a medium, and it excites a particular molecular degree of freedom, this excited degree of freedom, on average, has more energy, hence a higher temperature, than the rest of the medium. Although normal modes are in first approximation independent of each other, the remaining residual interactions tend to transfer the extra energy from the excited mode to other modes. This energy transfer may be hard to describe in exact terms, in particular because a number of different relaxation mechanisms may be present at the same time, but there is no doubt that all these mechanisms act in a somewhat friction-like manner and impede the vibrations of the excited mode. There is always a transfer of energy between different degrees of freedom in a medium. However, in equilibrium, each mode looses on average the same amount of energy as it receives. When a mode is excited by absorbing light passing through the medium, it receives additional energy and "warms up." The absolute temperature of that mode, T_a', becomes slightly higher than the temperature of the rest of the medium T_a. The energy that this mode transfers to other modes per unit time is now on average a bit larger than what it receives back from them. The end result is that the excited mode "cools down" and dissipates the absorbed energy to other modes in the medium. This process is not described by the HO model. The energy of a damped oscillator is reduced in proportion to the velocity of the mass that oscillates—not the difference between the temperatures of the excited mode and the medium. That implies that the precise shape of absorption bands as provided by the HO model may be different from the shape of real absorption bands as measured in experiments. Nevertheless, the end result is the same, that is, a peak of a certain height and width located at a particular wave number, and we rely on adjusting the three parameters of the HO model to model the shape of an actual peak.

3.4 LAMBERT LAW AND MOLECULAR CROSS SECTION

Lambert's law (Eq. 1.10) emerges in this formalism as a consequence of using a complex refractive index to describe the refractive index of the medium through which light propagates. Lambert's law is expressed in terms of the intensity of light $I(x)$. It says that as light propagates through a material, its intensity decays exponentially with the path length traveled through the medium. It is straightforward to convert this law into a differential form:

$$\frac{dI(x)}{dx} = -4\pi\kappa k\, I(x). \tag{3.7}$$

This form is equivalent to the integral form (Eq. 1.10) as one can easily see by verifying that Equation 1.10 is indeed a solution to Equation 3.7. The negative sign on the right-hand side of Equation 3.7 indicates that light intensity decreases with increasing path through the medium, as is expected. The constant of proportionality contains the absorption index and also the wave number of the incident light. The presence of the absorption index is fully expected. However, the presence of the wave number is somewhat unexpected. It says that, for the same value of the absorption index, a short wavelength light is absorbed more strongly than a long wavelength light.

A simple picture leading to the absorption law described by Equation 3.7 could be arrived at by the following reasoning. Imagine that each molecule in the absorbing medium has a cross section, σ, such that any photon impinging within the cross section σ is absorbed. Let us consider a cross-sectional area, A, perpendicular to the direction of propagation of light. In a time interval, dt, light travels a distance, dx. The number of photons that cross the area A is proportional to the product of light intensity $I(x)$ at the surface, the area A of the surface, and the distance dx. Let us assume that the medium contains N molecules per unit volume. The volume within which light travels is Adx and it thus contains $NAdx$ molecules. Each photon crossing the surface A has $\sigma NAdx/A$ probability of being absorbed because this represents the total absorbing area as a fraction of the total area A. These considerations lead to

$$\frac{dI(x)}{dx} = -N\sigma I(x), \tag{3.8}$$

which has the same form as Equation 3.7. There is a difference between the two equations. Equation 3.7 is exact, while Equation 3.8 is approximate in the sense that we ignored the possibility that some of the molecules could be positioned exactly behind the others, and therefore the effect of the molecules behind could be screened by the molecules in front of them. This would be less of a problem in a dilute medium with randomly positioned molecules than in a dense and/or highly ordered medium. With this said, the two expressions could be compared and we see that the implication is that the absorption index κ is proportional to the concentration N. These considerations support Beer's law. They also connect the classical electromagnetic theory that treats a medium as homogeneous with the molecular picture of a medium as being composed of molecules. The absorption of light by the medium is thus connected to the absorption of light by individual molecules. Aside from the expression (Eq. 3.8) being applicable for dilute media with randomly positioned molecules such as gas, the expression is independent of any particular model and, in a sense, could be seen as the definitional expression for the molecular cross section σ.

Another version of Lambert's law can be obtained by reexpressing it in terms of the energy density of the electromagnetic field $u(x)$. This can be achieved by noticing that $I(x) = c'u(x)$ and that for light propagating through a medium with speed $c' = c/n$, we have

$$dx = \frac{c}{n} dt. \tag{3.9}$$

Thus, Equation 3.7 can be rewritten as

$$\frac{du(x)}{dt} = -4\pi\kappa k \frac{c}{n} u(x). \tag{3.10}$$

The form (Eq. 3.10) is useful in those cases in which one wants to evaluate the energy absorbed per unit time in a particular volume of a sample. We will use the form (Eq. 3.10) later on when we turn to the calculation of absorbed energy per unit time associated with internal reflection.

4 Fresnel Equations

4.1 ELECTROMAGNETIC FIELDS AT THE INTERFACE

Just as Maxwell equations (Eq. 1.3) describe how light propagates through a medium, they also describe how light transmits through or reflects at an interface between two different media. Light is incident from medium 1 at the interface with medium 2. We assume that media 1 and 2 are isotropic and homogeneous, in which case we know that the solutions of Maxwell equations in such media are plane waves. The effects of the medium on light propagation are incorporated through the refractive index of the medium. Thus, for the incident electromagnetic wave, which propagates through medium 1, the electric field has the form

$$\mathbf{E}_{in}(\mathbf{x},t)=\mathbf{E}_{0in}e^{2\pi i n_1 \mathbf{k}_{in}\mathbf{x}-i\omega t}. \tag{4.1}$$

The electromagnetic theory does not tell us which waves are present in media 1 and 2; it just tells us the form of the solutions. Any linear superposition of waves of the form (Eq. 4.1) is also a solution. We have to specify how the incident wave travels through the medium and how it impinges on the interface. We also have to specify which waves we expect to have in the two media as a result of the presence of the interface. Obviously, if the interface is not present, the incident wave just keeps propagating through the medium. We could add additional electromagnetic waves of the form (Eq. 4.1) into the medium and they would all propagate through it without interacting one with another. Once we have two media and an interface between them, we have to add the waves that we expect we need in order to fully describe the situation. For instance, it is our experience that an interface both reflects and transmits an incident wave. When we look at the surface of a still shallow lake on a cloudy day, we see both the objects at the bottom as

Internal Reflection and ATR Spectroscopy, First Edition. Milan Milosevic.

well as the clouds above reflected by the surface of the lake. Thus, our experience tells us that we need reflected and transmitted waves and that the interface between the two media is what causes the incident wave to split into reflected and transmitted waves.

We have seen before that the electric field in an electromagnetic wave is accompanied by a magnetic field $\boldsymbol{B}_{in}$, oscillating with the same frequency and in phase with the electric field, and that Maxwell equations require that the electric and magnetic fields are mutually perpendicular and also perpendicular to the direction of motion determined by $\boldsymbol{k}_{in}$. The electric field vector and the wave vector of a plane wave define a plane in space. We refer to this plane as the plane of polarization of the wave.

Let the interface between media 1 and 2 be a flat plane as opposed to a curved surface. Light is incident from medium 1 onto the interface. The wave vector of incident light and the normal to the interface define a plane called the plane of incidence.

Let us pick the coordinate axes such that the z-axis is normal to the interface. For the x-axis, we pick the line in which the interface and the plane of incidence cross, and the y-axis is in the interface and is perpendicular to the plane of incidence. Thus, the interface is the plane $z = 0$, and the plane of incidence is the plane $y = 0$. The angle between the direction of propagation of incident light and the normal to the interface is defined as the angle of incidence θ. We call the angle between the direction of propagation of the reflected light and the normal to the interface the angle of reflection θ_R and, similarly, the angle of refraction φ for the transmitted (refracted) wave. We are not going to assume anything about the reflected and transmitted waves (except that they exist) and we will investigate what conditions on the reflected and transmitted light are imposed by the electromagnetic theory. Figure 4.1 illustrates the situation. Although the three waves in Figure 4.1 are shown all in one plane, we are not going to assume this.

The plane of polarization of the incident light can be arbitrary, but we will investigate two specific cases. In one case, we will assume that the plane of polarization of the incident light is the same as the plane of incidence; that is, the electric field vector oscillates in the plane of incidence. The incident light polarized in the plane of incidence is also called p-polarized light. In the other case, the plane of polarization of the incident light is perpendicular to the plane of incidence. The incident light polarized perpendicular to the plane of incidence is called s-polarized light. Obviously, the incident light polarized in an arbitrary direction could be decomposed into two waves, one p-polarized and the other one s-polarized. So by analyzing the two special cases, we end

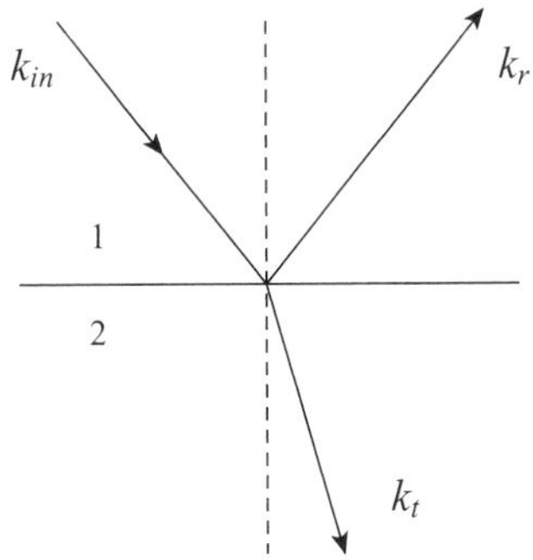

Figure 4.1 Incident, transmitted, and reflected light identified by the corresponding wave vectors at the interface between two media.

up with a formalism that allows us to analyze arbitrarily polarized incident waves. The general solutions for reflected and transmitted waves are of the following form (Eq. 4.1), that is,

$$\begin{aligned}\mathbf{E}_r(\mathbf{x},t) &= \mathbf{E}_{0r}e^{2\pi i n_1 \mathbf{k}_r \mathbf{x} - i\omega t}\\ \mathbf{E}_t(\mathbf{x},t) &= \mathbf{E}_{0t}e^{2\pi i n_2 \mathbf{k}_t \mathbf{x} - i\omega t}.\end{aligned} \tag{4.2}$$

Note that for the transmitted wave, we used the refractive index for medium 2. Also, vector **x** that marks the position in which the fields are followed is not to be mixed with the coordinate x as we defined it above. The coordinates of the point represented by vector **x** are (x, y, z).

4.2 SNELL'S LAW

The Maxwell equations for a medium give us quite a definite form for the transmitted and reflected waves. The only parameters not set in the above expressions are the magnitudes of the transmitted and reflected waves ($\mathbf{E}_{0r}$ and $\mathbf{E}_{0t}$) and the directions of the wave vectors $\boldsymbol{k}_r$ and $\boldsymbol{k}_t$ (magnitudes are known!). The three waves meet in the interface and that imposes on them further restrictions. The phase factors of all three waves must match in any point in the interface. That imposes

$$(n_1\boldsymbol{k}_{in}\boldsymbol{x})_{z=0} = (n_1\boldsymbol{k}_r\boldsymbol{x})_{z=0} = (n_1\boldsymbol{k}_t\boldsymbol{x})_{z=0}. \tag{4.3}$$

The requirement for all three phases to be the same will be further discussed later. Note that in Equation 4.3, we left out the terms in the phase factors that are already equal. Since **x** in Equation 4.3 is arbitrary and thus can be chosen perpendicular to $\boldsymbol{k}_{in}$, it follows that it must then also be perpendicular to $\boldsymbol{k}_r$ and $\boldsymbol{k}_t$. Thus, all three wave vectors are in the same plane—the plane of incidence. Equation 4.3 can be further reduced to

$$n_1 \sin\theta = n_1 \sin\theta_R = n_2 \sin\varphi. \tag{4.4}$$

The first part of the identity (Eq. 4.4) states that the angle of reflection equals the angle of incidence, that is, the well-known law of reflection. The second part of the identity is the well-known Snell law connecting the angle of refraction with the angle of incidence and the refractive indices of the two media. These are two foundational laws of geometrical optics. This is quite a bounty just from the requirement that the phases of the three waves must be equal in an arbitrary point in the interface. The laws of reflection and refraction tell us the angles of reflection and refraction, but they do not tell us how much of the incident light is reflected and how much is refracted; that is, they do not tell us the reflectance and transmittance of the interface.

4.3 BOUNDARY CONDITIONS AT THE INTERFACE

To find reflectance and transmittance of the interface, we have to use the boundary conditions that the Maxwell equations impose on the electric and magnetic fields at the interface. It is easy to see that these conditions require that the tangential component of the electric field is continuous at the interface and that the normal component of the electric field has a jump at the interface. This jump is such that the normal component of the electric field on one side of the interface multiplied by the dielectric constant of the medium is the same as the normal component of the electric field on the other side of the interface multiplied by the dielectric constant of the medium on the other side of the interface. Both normal and tangential components of the magnetic fields are continuous at the interface since we assume that the magnetic properties of optically relevant media are essentially the same as those of vacuum. This gives four equations. We can write them down by expressing the magnetic field of the plane wave in the form (Eq. 1.7) and by using a unit vector **n** normal to the interface. We recall that dot (scalar) product of any vector and **n** gives the normal component of that vector with respect to the interface, while the cross (vector) product of that vector with **n** gives tangential component of that vector to the interface. Thus, we have

$$
\begin{aligned}
\left[\varepsilon_1\left(\boldsymbol{E}_{in}+\boldsymbol{E}_r\right)-\varepsilon_2\boldsymbol{E}_t\right]\cdot\boldsymbol{n}&=0\\
\left[n_1\boldsymbol{k}_{in}\times\boldsymbol{E}_{in}+n_1\boldsymbol{k}_r\times\boldsymbol{E}_r-n_2\boldsymbol{k}_t\times\boldsymbol{E}_t\right]\cdot\boldsymbol{n}&=0\\
\left(\boldsymbol{E}_{in}+\boldsymbol{E}_r-\boldsymbol{E}_t\right)\times\boldsymbol{n}&=0\\
\left[n_1\boldsymbol{k}_{in}\times\boldsymbol{E}_{in}+n_1\boldsymbol{k}_r\times\boldsymbol{E}_r-n_2\boldsymbol{k}_t\times\boldsymbol{E}_t\right]\times\boldsymbol{n}&=0.
\end{aligned}
\tag{4.5}
$$

The system of Equation 4.5 looks a bit intimidating, but it is a compact way to write down the boundary conditions for the fields at the interface that we just described. Now we can further qualify our requirement (Eq. 4.3) that the phases of all the fields must be equal at the interface. It is apparent that the presence of the phase factors changing differently from point to point in the interface would make it impossible to have the boundary conditions (Eq. 4.5) satisfied in every point in the interface. Actually, the requirement that all three phases be equal is too strict and can be relaxed to allow the phases to differ for a constant value independent of the position on the interface. The phases change from a point to a point in the interface, but they change together and can thus be divided out from the expressions (Eq. 4.5).

4.4 FRESNEL FORMULAE

Now it should become apparent why we separated the analysis into two different cases with respect to the polarization of the incident wave. For the s-polarized light, all the electric fields are perpendicular to the plane of incidence, so the first equation in Equation 4.5 is identity. The second equation in Equation 4.5 yields Snell's law. The third and fourth equations give

$$
\begin{aligned}
E_{in}+E_r-E_t&=0\\
n_1\left(E_{in}-E_r\right)\cos\theta-n_2E_t\cos\varphi&=0
\end{aligned}
\tag{4.6}
$$

These two equations can be solved in terms of the ratios r^s and t^s, defined as

$$
\begin{aligned}
r^s&=\frac{E_r}{E_{in}}\\
t^s&=\frac{E_t}{E_{in}},
\end{aligned}
\tag{4.7}
$$

where superscript s identifies the s-polarized incident light. We find

$$\begin{aligned} r^s &= \frac{n_1 \cos\theta - \sqrt{n_2^2 - n_1^2 \sin^2\theta}}{n_1 \cos\theta + \sqrt{n_2^2 - n_1^2 \sin^2\theta}} \\ t^s &= \frac{2n_1 \cos\theta}{n_1 \cos\theta + \sqrt{n_2^2 - n_1^2 \sin^2\theta}}. \end{aligned} \tag{4.8}$$

A similar procedure yields for the p-polarized incident light

$$\begin{aligned} (E_{in} - E_r)\cos\theta - E_t \cos\varphi &= 0 \\ n_1 (E_{in} + E_r) - n_2 E_t &= 0, \end{aligned} \tag{4.9}$$

which, for the ratios r^p and t^p defined in analogy with Equation 4.7, gives

$$\begin{aligned} r^p &= -\frac{n_2^2 \cos\theta - n_1 \sqrt{n_2^2 - n_1^2 \sin^2\theta}}{n_2^2 \cos\theta + n_1 \sqrt{n_2^2 - n_1^2 \sin^2\theta}} \\ t^p &= \frac{2n_1 n_2 \cos\theta}{n_2^2 \cos\theta + n_1 \sqrt{n_2^2 - n_1^2 \sin^2\theta}}. \end{aligned} \tag{4.10}$$

Equations 4.8 and 4.10 are the well-known Fresnel equations. They provide the magnitudes of the electric fields of reflected and transmitted waves in terms of the magnitude of the electric field of the incident wave.

4.5 REFLECTANCE AND TRANSMITANCE OF INTERFACE

We are interested in finding the reflectance and transmittance of the interface. To calculate the reflected and the transmitted power, we need to know both the beam intensity and the cross section of the beam. For simplicity, we assume the intensity is uniform over the cross section of the beam.

Also, we see that, because the angle of reflection is the same as the angle of incidence, the cross section of the reflected beam remains the same as that of the incident beam. However, since the angle of refraction is different from the angle of incidence, the cross section of the transmitted beam C_2 (Fig. 4.2) is different from the cross section of the incident beam C_1. From Figure 4.2, it follows that the ratio of the two cross sections is

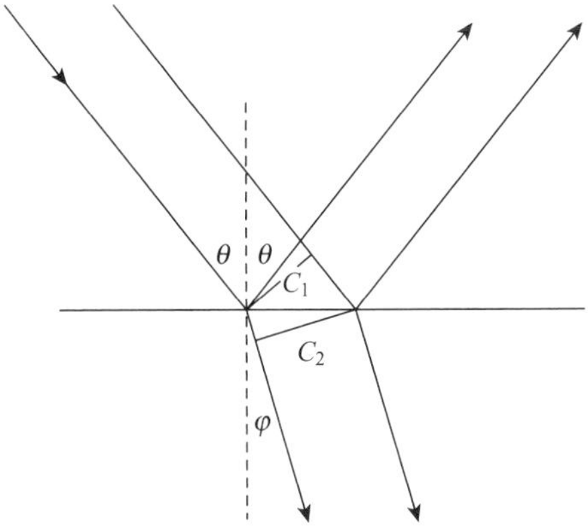

Figure 4.2 Change of beam's cross section on refraction.

$$\frac{C_1}{C_2} = \frac{\cos\theta}{\cos\varphi}. \tag{4.11}$$

Beam intensity, as we have seen before, is simply the energy density of the electromagnetic field multiplied by the speed of light. However, the results (Eqs. 3.2 and 3.3) we found earlier were for the electromagnetic wave in vacuum. These results must be modified in the presence of a medium. The speed of light in the medium is smaller than in vacuum; that is, it is c/n. For the energy density (Eq. 3.2), the modification is

$$u = \frac{\varepsilon}{4\pi} E^2; \tag{4.12}$$

that is, the free space expression for energy density is multiplied by the dielectric constant of the medium. The dielectric constant of a medium is refractive index squared. Thus, the power of the incoming beam is

$$P_{in} = u_{in} \frac{c}{n_2} C_1 = \frac{E_{in}^2}{4\pi} cn_1 C_1. \tag{4.13}$$

Reflected and transmitted powers are

$$P_r = \frac{E_r^2}{4\pi} cn_1 C_1 \tag{4.14a}$$

$$P_t = \frac{E_t^2}{4\pi} cn_2 C_2. \tag{4.14b}$$

Thus, the reflectance and transmittance of the interface are

$$R = \frac{P_r}{P_{in}} = \frac{E_r^2}{E_{in}^2} = |r|^2 \tag{4.15}$$

and

$$T = \frac{P_t}{P_{in}} = \frac{E_t^2}{E_{in}^2}\frac{n_2 C_2}{n_1 C_1} = |t|^2 \frac{\sqrt{n_2^2 - n_1^2 \sin^2\theta}}{n_1 \cos\theta}. \tag{4.16}$$

Note that we are assuming that the electric field magnitudes in the above expressions are real numbers. However, we are using a complex form for the plane waves in Equations 4.1 and 4.2, and where these complex forms are used, we expect that the absolute values of those expressions are used for the fields in the above equations. This is mirrored in the expressions above by the use of the absolute value of the Fresnel amplitude coefficients. The coefficients themselves may be complex, signifying that the fields are shifted in phase. As we have seen before, this shift is allowed as long as it is the same everywhere in the interface.

It is easy to verify by direct substitution that $R + T = 1$ for both polarizations. Note that this result simply restates the conservation of energy. We didn't have to postulate that energy is conserved; the result simply emerged from electromagnetic theory. At the interface, the electromagnetic waves are either reflected or transmitted. No electromagnetic energy is absorbed in the interface regardless whether the media themselves are transparent or whether they absorb light.

4.6 SNELL'S PAIRS

Snell's law connects the angles of incidence and refraction. The two angles form a pair (θ, φ). Light is incident from medium 1 at angle θ and refracts at the interface into medium 2 at angle φ. The property of such a pair is that if the direction of light is reversed, and thus φ becomes the angle of incidence, the angle of refraction that will be given by Snell's law will be θ; that is, the roles of the two angles reverse. The situation is not fully symmetric since the reflected ray is always in the incident medium.

We can consider the reflectance and transmittance amplitude coefficients r and t as functions of the Snell pair (θ, φ). The first angle in the pair is the angle of incidence. The second angle is not an independent variable since it is given by Snell's law, so we cannot put an arbitrary angle here, only the Snell counterpart of the angle of incidence. However, we can reverse the angles of the pair since they are each other's counterpart. The question is how are $r(\theta, \varphi)$ and $r(\varphi, \theta)$ connected, if at all? Since n_2 and φ have to replace n_1 and θ and vice versa, an inspection of Equations 4.8 and 4.10 shows that the reflectance coefficients simply reverse signs. Transmission coefficients, however, change strengths. This is an interesting symmetry of the reflection coefficients that we will use later on. Note here that reflection amplitude coefficients are more relevant to us than transmission coefficients since, in spectroscopy, we are very likely to encounter reflectance from a single interface. The detector and the source are virtually always in air (or vacuum), so whatever medium light enters from air, it eventually has to return back to air in order to reach the detector. So, unlike the single interface reflectance coefficients, the single interface transmittance coefficients are rarely the subject of spectroscopic measurements.

4.7 NORMAL INCIDENCE

For normal incidence, Fresnel Equations 4.8 and 4.10 take the form

$$r = \frac{n_1 - n_2}{n_1 + n_2} \tag{4.17a}$$

$$t = \frac{2n_1}{n_1 + n_2}. \tag{4.17b}$$

The expressions are the same for both polarizations. The plane of incidence is defined by the normal to the interface and the direction of the incident wave. For normal incidence, these two directions coincide, making it impossible to define the plane of incidence and thus impossible to distinguish between the s and p polarizations.

4.8 BREWSTER'S ANGLE

Another important special case is the one of Brewster's angle. By direct inspection of Equations 4.8 and 4.10, we can see that r^s cannot be zero for any angle of incidence. However, r^p vanishes when the angle of incidence θ satisfies

$$\sin\theta = \frac{n_1}{\sqrt{n_1^2 + n_2^2}}. \tag{4.18}$$

This angle is referred to as Brewster's angle. Since the reflectance of the interface for *p*-polarized light at Brewster's angle is zero, it is also referred to as the polarizing angle. The reason for this is that if unpolarized light incident at Brewster's angle strikes the interface, only the *s*-polarized component is reflected. The *p*-polarized component is completely transmitted through the interface. Thus, by reflecting unpolarized light from an interface at Brewster's angle, we can produce light that is linearly polarized. Generally, only a small fraction of the incident light intensity is reflected, so this method of polarizing light is not very efficient. Light transmitted through the interface contains all the *p*-polarized light that was in the incident light but is missing a portion of the *s*-polarized component that was reflected at the interface. So, the transmitted light is partially polarized. However, the transmitted light can be made incident on another interface at Brewster's angle to reflect off more of the *s*-polarized component. No *p*-polarized component is lost in the process, so this can be repeated until only a negligible intensity of the *s*-polarized component remains in the transmitted beam. Note that while each reflected *s*-polarized component is weak, it is perfectly polarized. On the other hand, the *p*-polarized component is strong, but it requires multiple transmissions through a number of interfaces until the transmitted component is sufficiently purged of the *s*-polarized component for the transmitted light to be *p*-polarized.

4.9 THE CASE OF THE 45° ANGLE OF INCIDENCE

Another important case is the one in which the angle of incidence is 45°. For θ=45°, the Fresnel equations simplify to

$$r^p = -\frac{n_2^2 - n_1\sqrt{2n_2^2 - n_1^2}}{n_2^2 + n_1\sqrt{2n_2^2 - n_1^2}}$$

$$r^s = -\frac{n_1 - \sqrt{2n_2^2 - n_1^2}}{n_1 + \sqrt{2n_2^2 - n_1^2}}.$$

It can be seen by direct inspection that the square of r^s equals r^p up to the sign. This translates into the reflectance of an interface for *p*-polarized light being equal to the square of the reflectance for the

s-polarized light regardless of the refractive indices of the two media. This is unusual, but the exact result will be used later.

4.10 TOTAL INTERNAL REFLECTION

The reflectance of a typical interface for both s and p polarizations are shown in Figure 4.3.

The interface of Figure 4.3 is between vacuum (or air) with refractive index $n_1=1$, and a typical transparent medium with the refractive index $n_2=1.5$. Light is incident from the vacuum side. The dip in reflectance at Brewster's angle around 55° is clearly visible. The reflectance for both polarizations is the same (about 4%) at normal incidence. Both polarizations reflect totally at grazing angle $\theta = 90°$. The general features of the reflectance curves shown in Figure 1.1 are present for a wide variety of refractive indices as long as the incident side is in the lower refractive index medium.

What happens if light is incident from the higher refractive index side? We have already seen that the angle of incidence and reflection form a pair connected by Snell's law. Every angle of incidence θ on the smaller refractive index side is accompanied by an angle of refraction φ on the larger refractive index side and the reflectance of the interface

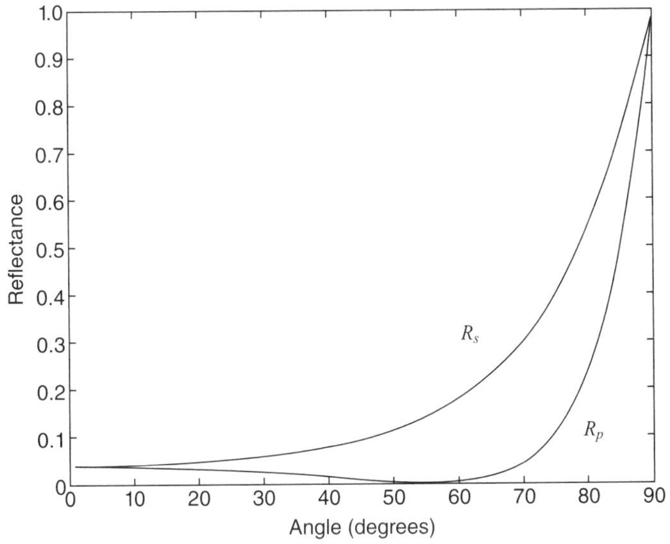

Figure 4.3 R_s and R_p for an interface as functions of the angle of incidence.

remains the same if the roles of θ and φ are reversed. If light is incident from the optically rarer medium, the angle of refraction is smaller than the angle of incidence. For the largest possible angle of incidence $\theta=90°$, the angle of refraction φ_{max} is some angle smaller than 90°. This maximum angle of refraction is given by Snell's law as

$$\sin(\varphi_{max}) = \frac{n_1}{n_2}. \tag{4.19}$$

Thus, starting from normal incidence, where both angles are zero, the angle of incidence grows quicker than the angle of refraction, and as the angle of incidence approaches its highest possible value of 90°, the angle of refraction approaches φ_{max}. If we reverse the roles of the angles of incidence and refraction so light is now incident from the higher refractive index side, and we start with normal incidence, the angle of refraction is now always larger than the angle of incidence, as Snell's law requires. As the angle of incidence approaches φ_{max}, the angle of refraction approaches 90°. It is always tricky to reverse the roles of the angles of incidence and refraction. Let us therefore proceed carefully by showing in Figure 4.4 the diagrams of the two cases.

We have to be careful not to confuse and intermix the notation, which can be easily done by reversing the direction of light. So let us upgrade our notation for the amplitude coefficients by introducing subscripts that denote the media of the interface by specifying first the incident medium and second the transmitted medium. For instance, the coefficient r_{12}^s indicates the reflectance amplitude coefficient for the s-polarized light incident from medium 1 onto the interface with medium 2. Accordingly, the expression for the reflectance coefficient r^s could be written as

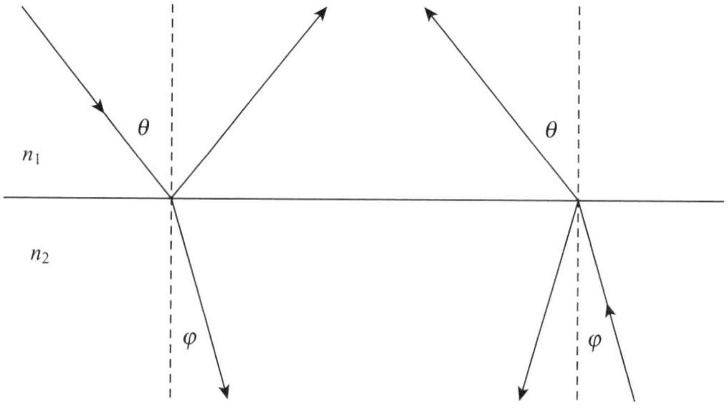

Figure 4.4 Reversal of the direction of light.

$$r_{ab}^{s} = \frac{n_a \cos\alpha - \sqrt{n_b^2 - n_a^2 \sin^2\alpha}}{n_a \cos\alpha + \sqrt{n_b^2 - n_a^2 \sin^2\alpha}}. \tag{4.20}$$

Here, subscript a denotes the incident and subscript b the transmitted medium. Angle α is the angle of incidence, and it is by definition always in medium a. Thus, the left side of Figure 4.4 corresponds to the case in which $a=1$, $b=2$, $\alpha=\theta$. Reversing the beam direction corresponds to setting $a=2$, $b=1$, $\alpha=\varphi$. Thus, we would have for the two cases

$$r_{12}^{s} = \frac{n_1 \cos\theta - \sqrt{n_2^2 - n_1^2 \sin^2\theta}}{n_1 \cos\theta + \sqrt{n_2^2 - n_1^2 \sin^2\theta}} \tag{4.21a}$$

$$r_{21}^{s} = \frac{n_2 \cos\varphi - \sqrt{n_1^2 - n_2^2 \sin^2\varphi}}{n_2 \cos\theta + \sqrt{n_1^2 - n_2^2 \sin^2\varphi}}, \tag{4.21b}$$

and, similarly, for the p-polarization,

$$r_{12}^{p} = -\frac{n_2^2 \cos\theta - n_1\sqrt{n_2^2 - n_1^2 \sin^2\theta}}{n_2^2 \cos\theta + n_1\sqrt{n_2^2 - n_1^2 \sin^2\theta}} \tag{4.21c}$$

$$r_{21}^{p} = -\frac{n_1^2 \cos\varphi - n_2\sqrt{n_1^2 - n_2^2 \sin^2\varphi}}{n_1^2 \cos\varphi + n_2\sqrt{n_1^2 - n_2^2 \sin^2\varphi}}. \tag{4.21d}$$

Note that in the reverse case, the angle of incidence is φ, while the angle of refraction is θ. The two angles are a Snell pair. The angle φ can be any angle in the range 0 to $\varphi_{\max}$. As φ sweeps the range 0 to $\varphi_{\max}$, θ sweeps the range from 0 to 90°. But there is nothing to prevent us from increasing the angle φ beyond $\varphi_{\max}$. So what happens if the angle of incidence continues to increase past the maximum angle $\varphi_{\max}$? Obviously, the angle of refraction cannot exceed 90°. If the angle of incidence φ exceeds $\varphi_{\max}$, we get the angle of refraction θ for which $\sin\theta > 1$. This, of course, is impossible for any real angle. Thus, the refracted ray can no longer physically exist because there is no angle for it to refract into.

Let us see how that is reflected in the Fresnel amplitude coefficients. Our expressions for amplitude coefficients do not include the angle of refraction. We expressed everything in terms of the angles of incidence. To achieve that, we used the Snell law. The Snell law enabled the expressions for amplitude coefficients to remain meaningful even in the cases where refraction no longer occurred. However, as we extend the angle of incidence beyond the maximum φ_{max}, a new problem arises. The expression under the square root becomes negative. Hence, the square root becomes imaginary. Both reflection amplitude coefficients now have the form

$$r = \frac{A - iB}{A + iB}. \tag{4.22}$$

Thus, the reflectance amplitude coefficients become complex numbers. This, however, is not necessarily a problem. The reflectance calculated from Equation 4.22 by taking the square of absolute value is

$$R = |r|^2 = \frac{A^2 + B^2}{A^2 + B^2} = 1. \tag{4.23}$$

The result (Eq. 4.23) holds for both s and p polarizations. Thus, the reflectance of the interface becomes total when the angle of incidence exceeds the critical angle defined as the angle for which the square root term in the Fresnel equations $\sqrt{n_1^2 - n_2^2 \sin^2 \varphi}$ becomes zero. The critical angle is thus

$$\sin \varphi_c = \frac{n_1}{n_2}, \tag{4.24}$$

which is, of course the same as the maximum angle of refraction (Eq. 4.19) for the case of reverse beam direction. What is the meaning of total reflection? Generally, incident light at the interface splits into two components, one reflected and one refracted. The power of the incident beam is thus split so that the sum of the transmitted and reflected power adds to the incident power. However, for internal reflection defined as the reflection at an interface for which the incident medium has a higher refractive index than the transmitted medium, there is a critical angle of incidence for which there is no transmitted beam so the entire power of the incident beam flows into the reflected beam. This is mirrored by the fact that now, the reflectance (Eq. 4.23) of the interface is total (100%). The reflectance amplitude coefficients are complex numbers of unit absolute value. Such numbers can be written as

$$r = e^{i\psi}, \tag{4.25}$$

where ψ is a real number. The meaning of ψ is that the reflection has added a constant phase shift to the reflected wave. In the notation of Equation 4.22, the phase shift is

$$\tan\psi = \frac{2AB}{A^2 + B^2}. \tag{4.26}$$

The phase shift is the same everywhere in the interface. This is allowed by the requirements (Eq. 4.3). Also, the phase shift is different for the two polarizations.

When light is incident from air onto an interface with an optical medium, the reflectance occurring at the interface is called external. If, on the other hand, light is incident from a medium onto an interface with air, the reflection is called internal. We can generalize this naming convention to refer to external reflection whenever light is incident from a medium with a lower refractive index onto an interface with a medium with a higher refractive index. If light is incident from a medium with a higher refractive index onto a medium with a lower refractive index, we call the reflection internal. The real distinction is that for internal reflection, there is a range of angles of incidence for which the reflection is total. Total reflection is different from the reflection from a very good mirror. Although a mirror reflection can be very high, there is always a fraction of light intensity that is absorbed during the reflection. If light reflects multiple times from the mirror, a loss in intensity quickly becomes apparent. Even if special coatings are applied over the mirror surface to enhance the reflectivity and the reflectance of the surface approaches total reflectance, the effect is confined to a limited spectral range. Total internal reflectance, on the other hand, is truly total. In optical fibers used for telecommunications, light internally reflects thousands of times and travels for miles.

The reflectance for two polarizations of incident light for internal reflection is the absolute value squared of the amplitude coefficients (Eq. 4.21b,d). The internal reflection counterpart of the external reflection shown in Figure 4.3 is displayed in Figure 4.5. We clearly see the split of the internal reflection phenomenon into two distinct ranges of angle of incidence. One regime of internal reflection occurs for angles of incidence smaller than the critical angle. In this regime, the reflectance curves resemble the reflectance curves of the external reflection case except that the angular scale is now squeezed into the 0–φ_c range. Actually, more than just a resemblance, the reflectance for the internal reflection at an angle of incidence φ is exactly the same as the reflectance for the angle of incidence θ for the associated external reflection,

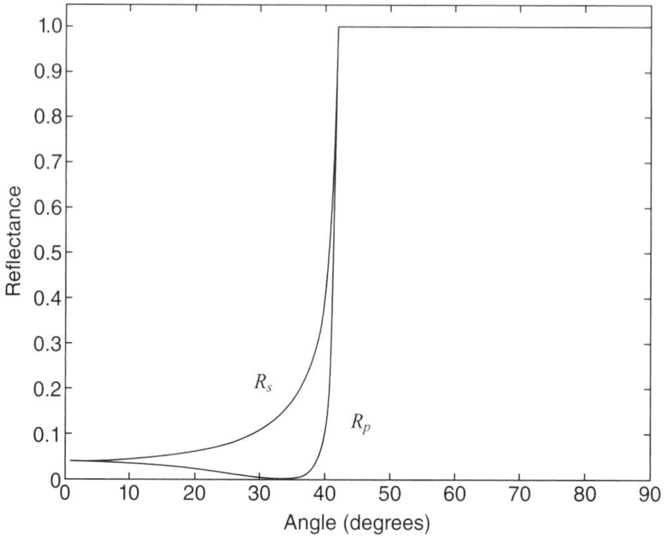

Figure 4.5 The internal reflectances R_s and R_p versus the angle of incidence.

where φ and θ are angles that form a Snell pair. This is true for both polarizations.

We call this regime of internal reflection subcritical internal reflection because the angle of incidence is smaller than critical. We call the regime of internal reflection for angles of incidence above the critical supercritical internal reflection. For the case where the refractive indices of the two media are real, that is, the two media do not absorb light, the supercritical internal reflection is total.

5 Evanescent Wave

5.1 EXPONENTIAL DECAY AND PENETRATION DEPTH

For subcritical internal reflection, the waves present on the two sides of the interface are still incident, reflected, and transmitted just as for external reflection. However, as we have seen, this situation changes for supercritical internal reflection. Only incident and reflected waves are present. So what happened to the transmitted wave? The electromagnetic fields must be present on both sides of the interface as the boundary conditions require. Let us examine the solution (Eq. 4.2) for the transmitted wave. The factor of interest is the exponent of the exponential function. It contains the scalar product of the wave vector of the transmitted wave and the radius vector of a point of observation. We assume that the plane of incidence is the x-z plane with the x-axis in the interface and the z-axis normal to the interface. The scalar product then has two components:

$$in_2\boldsymbol{k}_t\boldsymbol{x} = in_2(k_x x + k_z z) = in_2 k(x\sin\varphi + z\cos\varphi). \tag{5.1}$$

Remember that, for supercritical internal reflection, there is really no angle of refraction φ. Still, we can use Snell's law to replace the angle of refraction in Equation 5.1 with the angle of incidence. The result is

$$\begin{aligned} in_2 k(x\sin\varphi + z\cos\varphi) &= ik\left(xn_1\sin\theta + z\sqrt{n_2^2 - n_1^2\sin^2\theta}\right) \\ &= ikxn_1\sin\theta - kz\sqrt{n_1^2\sin^2\theta - n_2^2}. \end{aligned} \tag{5.2}$$

Inserting Equation 5.2 into the expression for the transmitted wave leads to

$$\boldsymbol{E}_t(\boldsymbol{x},t) = \boldsymbol{E}_{0t}e^{-i\omega t}e^{2\pi i n_1 kx\sin\theta}e^{-2\pi kz}\sqrt{n_1^2\sin^2\theta - n_2^2}. \tag{5.3}$$

Internal Reflection and ATR Spectroscopy, First Edition. Milan Milosevic.

The first exponential term on the right-hand side of Equation 4.3 is the standard oscillatory term that expresses the time dependence of the wave. The second term is just a standard term describing the propagation of the wave along the interface. These two terms look exactly the same as they look in the case of subcritical internal reflection. The difference is that the term usually describing the propagation along the z-axis is no longer describing any propagation. Instead, it is describing the exponential decay of the amplitude of the wave propagating along the interface. The wave is confined to the interface. The amplitude of the wave is the largest at the interface and is exponentially damped into the rarer medium with the distance away from interface. This surface wave is called the evanescent wave, and it is a remnant of the refracted wave that lingers in the rarer medium as the angle of incidence crosses from the subcritical into the supercritical regime of internal reflection.

The amplitude of the evanescent wave decays to $1/e$ of its maximum value at a distance d_p from the interface. This distance, called the penetration depth, is on the order of the wavelength of light. From Equation 4.2, it can be seen that the expression for the penetration depth is

$$d_p = \frac{1}{2\pi k\sqrt{n_1^2 \sin^2\theta - n_2^2}} = \frac{\lambda}{2\pi\sqrt{n_1^2 \sin^2\theta - n_2^2}}. \tag{5.4}$$

The penetration depth of the evanescent wave is a function of the refractive indices of the two media and the angle of incidence. The penetration depth is defined only for supercritical internal reflection. Below the critical angle, the expression for the penetration depth is imaginary. Also, the penetration depth is the same for both polarizations of incident light.

It is interesting that the penetration depth as a function of the angle of incidence decreases from infinite value at the critical angle down to its minimum value at grazing incidence (θ=90°).

The dependence of the penetration depth on the angle of incidence for the incident light with the wavelength of 10 μm (i.e., the wave number of 1000 cm^{-1}) is illustrated in Figure 5.1.

It can be seen from Equation 5.3 that the evanescent wave propagates parallel to the interface with the velocity c' given by

$$c' = \frac{c}{n_1 \sin\theta}. \tag{5.5}$$

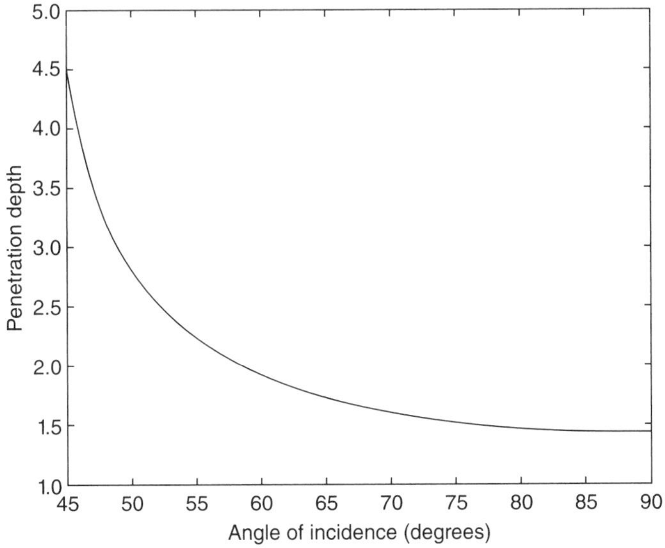

Figure 5.1 Penetration depth (in micrometers) versus angle of incidence with refractive indices $n_1=1.5$ and $n_2=1$.

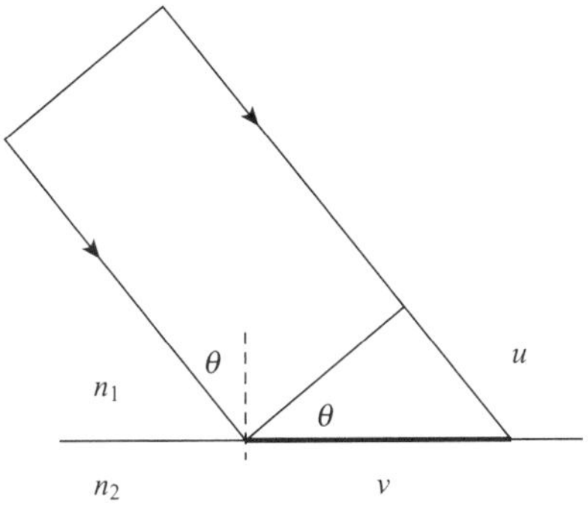

Figure 5.2 Speed of the evanescent wave.

Since $\sin\theta$ is always less than one, the speed of propagation of the evanescent wave is always larger than the speed of light in the incident medium. As the wave front of the incident wave travels toward the interface, it is inclined to the interface at an angle of $\pi/2 - \theta$ so as the successive points on the wave front reach the interface, they must always meet the same phase of the traveling evanescent wave. The situation is sketched in Figure 5.2. By the time the incident light traverses the distance u traveling at speed c/n_1, the evanescent wave must traverse the distance v and since

$$v = \frac{u}{\sin\theta}.$$

The speed given by Equation 5.5 achieves that.

Since the evanescent wave (Eq. 5.3) propagates parallel to the interface, the energy flow contained in the evanescent wave also travels parallel to the interface. There is no wave propagation perpendicular to the interface and thus there is no energy flow through the interface. We will revisit and revise this conclusion later.

5.2 ENERGY FLOW AT A TOTALLY INTERNALLY REFLECTING INTERFACE

A round beam of light of diameter D incident onto a totally internally reflecting interface illuminates an elliptically shaped spot on the interface. The minor diameter of the elliptical spot is the same as the diameter D of the incident beam, and the major diameter is equal to $D/\cos\theta$. The evanescent wave is induced only on this illuminated area of the interface as indicated in Figure 5.3. The rest of the interface that is not illuminated by the incident beam does not contain the evanescent wave. Thus, on the edge of the illuminated spot, there exists a transitional area in which the evanescent wave decays from full strength down to zero strength. The width of this transitional region must be on the order of the wavelength of incident light.

Most textbooks covering the electromagnetic theory mention total internal reflection as a curiosity. If they discuss the evanescent wave at all, the discussion typically centers on the description of the evanescent

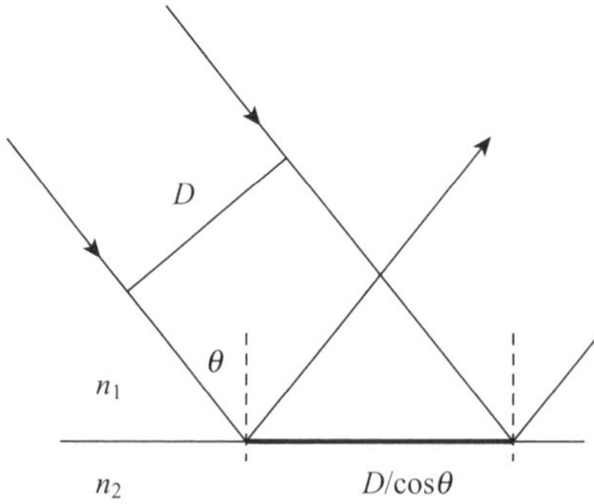

Figure 5.3 Area of totally reflecting interface containing an evanescent wave.

wave given by the expression (Eq. 5.3) and conclude that the wave propagates parallel to the interface and that there is no net flow of energy through the interface. Our description of total internal reflection of electromagnetic waves was based on plane wave solutions and, since plane waves are infinite in extent, does not apply to a situation where a finite area of the interface is illuminated. Therefore, the results we have derived do not apply to the boundary region itself. Thus, it would be natural to expect that the boundary of the illuminated region is responsible for transferring the electromagnetic energy through the interface and into and out of the evanescent wave. We will come back to the issue of energy transport through the interface to and from the evanescent wave when we analyze attenuated total reflection (ATR) spectroscopy of powders. As we will see, this will change the conventional description of the evanescent wave.

5.3 THE EVANESCENT WAVE IN ABSORBING MATERIALS

It would make little sense to use an absorbing medium as the incident medium for total internal reflection. Consequently, we will always assume that the incident medium (i.e., ATR material) is fully transparent for the incident radiation. As we have seen, that translates into requiring that the refractive index of the incident material is a real number. Also, the incident material will have to have a high index of refraction, n_1, so that the conditions for total internal reflection can be achieved with a wide variety of samples that are brought into intimate contact with the incident material. However, the second medium could generally be an absorbing medium with a nonvanishing absorption index. The refractive index of a typical material (water solutions, polymers, hydrocarbons, lipids, etc.) is around 1.5. These materials strongly absorb infrared light. Let us see what the effect of an absorbing rarer medium is on the evanescent wave (Eq. 5.3). We can, according to Equation 1.8, write the refractive index of the absorbing rarer medium as $n_2 = n + i\kappa$. Inserting this into Equation 5.3 yields

$$\boldsymbol{E}_t(\boldsymbol{x},t) = \boldsymbol{E}_{0t} e^{-i\omega t} e^{2\pi i n_1 k x \sin\theta} e^{-2\pi k z \sqrt{n_1^2 \sin^2\theta - n^2 + \kappa^2 - 2in\kappa}}. \tag{5.6}$$

The interesting part of the above expression is the last exponential term containing the z-axis dependence. The square root term is real for the nonabsorbing rarer medium. For the absorbing rarer medium, the square root becomes complex. Writing

$$\sqrt{n_1^2 \sin^2 \theta - n^2 + \kappa^2 - 2in\kappa} = U + iV, \tag{5.7}$$

we can rewrite the z-axis dependent term as

$$e^{-2\pi kzU - 2\pi ikzV}. \tag{5.8}$$

Thus, the term describing the z-axis dependence acquires a propagation factor along the z-axis. That means that now electromagnetic energy leaks through the interface and propagates into the absorbing medium. The interface in itself does not actually absorb any electromagnetic energy. The absorbing rarer medium makes the interface slightly "transparent." The electromagnetic energy leaks through the interface and into the absorbing medium where it is ultimately absorbed. The evanescent wave no longer propagates parallel to the interface. It now propagates at a slight angle to the interface. The energy that the evanescent wave loses must be constantly replenished by leaking some of the incident electromagnetic energy through the interface. Thus, the reflected wave no longer carries all the incident energy and, consequently, the reflection is no longer total. The reflection is said to be attenuated. This means that the reflectance of such an interface is no longer one (i.e., 100%) but is less than one. This phenomenon has great practical utility in spectroscopy and is known as ATR.

We can see from Equations 5.6–5.8 that the attenuation is directly caused by the absorption index. If the absorption index is zero, V is zero and the reflection is total. For a nonzero absorption index, the reflection is attenuated. To see this more explicitly, we can calculate U and V from Equation 5.7 by taking square of both sides and equating real and imaginary parts. To simplify the expressions, we limit ourselves to the important case of the absorption index sufficiently smaller than one so that we can disregard the terms containing κ^2 and higher terms. The result is

$$V = -\frac{n\kappa}{\sqrt{n_1^2 \sin^2 \theta - n^2}} \tag{5.9}$$

and

$$U = \sqrt{n_1^2 \sin^2 \theta - n^2}. \tag{5.10}$$

As expected, the term that controls the "leakage" of the incident energy through the interface (Eq. 5.9) is proportional to the absorption index.

6 Electric Fields at a Totally Internally Reflecting Interface

6.1 E_X, E_Y, AND E_Z FOR S-POLARIZED INCIDENT LIGHT

To find electric fields in all three spatial directions at a totally internally reflecting interface, we recall our choice of the coordinate system. The interface is in the x-y plane. The z-axis is thus perpendicular to the interface. The plane of incidence is the x-z plane. Again we split the problem into two cases: the case where the incident wave is polarized in the plane of incidence and the case where the incident wave is polarized perpendicular to the plane of incidence. The incident wave polarized in the plane of incidence, or p-polarized, has the electric field in the plane of incidence so the electric field has only x- and z-components. The incident wave polarized perpendicular to the plane of incidence, or s-polarized, has the electric field perpendicular to the plane of incidence so the electric field has only a y-component. For the s-polarized incident light, the electric fields of the incident, the reflected and the transmitted light are all perpendicular to the plane of incidence.

Using Fresnel equations, it is easy to express the electric field of the transmitted wave:

$$E_t = E_0 t_{12} = \frac{2n_1 \cos\theta}{n_1 \cos\theta + \sqrt{n_2^2 - n_1^2 \sin^2\theta}} E_0. \tag{6.1}$$

Note that the transmitted wave becomes the evanescent wave for the angle of incidence θ larger than critical and that Equation 6.1 can then be written as

$$E_t = \frac{2n_1 \cos\theta}{n_1 \cos\theta + i\sqrt{n_1^2 \sin^2\theta - n_2^2}} E_0.$$

Internal Reflection and ATR Spectroscopy, First Edition. Milan Milosevic.
© 2012 John Wiley & Sons, Inc. Published 2012 by John Wiley & Sons, Inc.

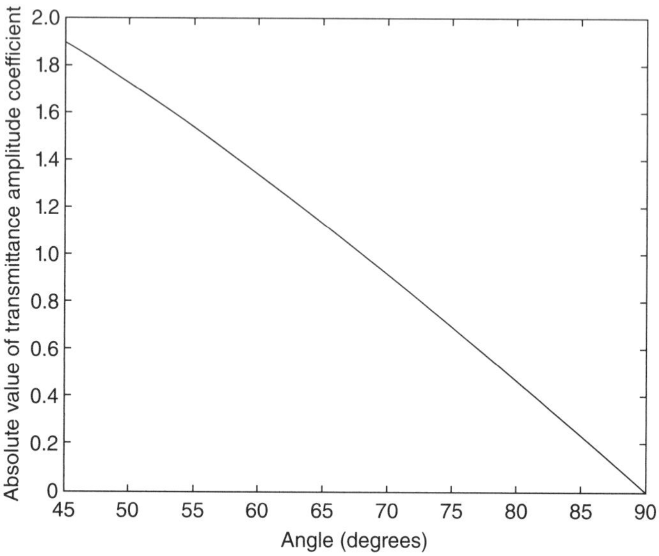

Figure 6.1 The electric field of the evanescent wave in units of the amplitude of the incident field versus the angle of incidence. The refractive index of the incident medium is $n_1 = 1.5$ and the refractive index of the rarer medium is $n_2 = 1$.

The explicit appearance of the imaginary unit in the denominator indicates that the electric field of the evanescent wave is out of phase with the incident wave. Figure 6.1 shows the amplitude of the electric field of the evanescent wave expressed in terms of the amplitude of the electric field of the incident wave for a typical medium. It is interesting that the electric field amplitude of the evanescent wave is stronger than the electric field amplitude of the incident wave for the angles near the critical angle and vanishes near the grazing incidence. Exactly at the critical angle, the electric field amplitude of the incident wave is exactly twice the strength of the electric field amplitude of the incident wave regardless of the refractive indices of the two media. Thus, for the *s*-polarized light, the components of the electric field of the evanescent wave are

$$E_x = 0, E_y = \frac{2n_1 \cos\theta}{n_1 \cos\theta + \sqrt{n_2^2 - n_1^2 \sin^2\theta}} E_0, E_z = 0. \tag{6.2}$$

6.2 E_X, E_Y, AND E_Z FOR *P*-POLARIZED INCIDENT LIGHT

For the *p*-polarized incident wave, the situation is slightly more complicated because the electric fields of the incident, reflected and transmitted waves are not collinear. The situation is depicted in Figure 6.2.

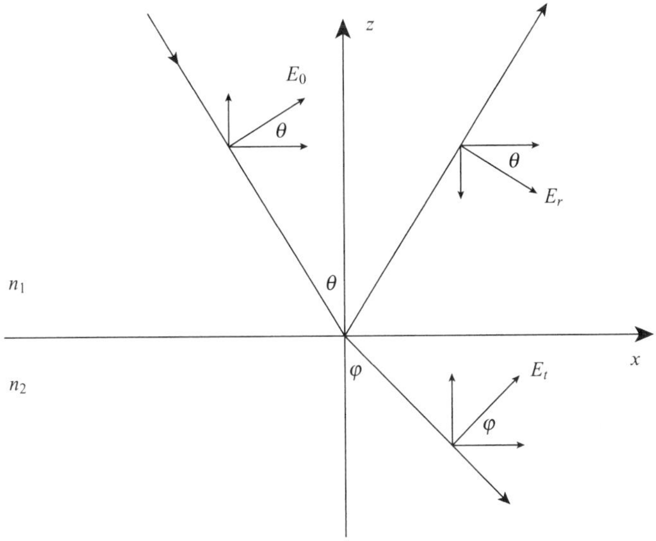

Figure 6.2 Electric fields at an interface for p-polarized incident wave.

The electric field of the transmitted wave has components along the x- and z-axes as follows:

$$E_x = E_t \cos\varphi,\ E_z = -E_t \sin\varphi. \tag{6.3}$$

In Equation 6.3, we used the case of subcritical internal reflection to express the components of the electric field of the transmitted wave in terms of the angle of refraction. Now we can use Snell's law to replace the angle of refraction φ with the incident angle θ and Fresnel equations for p-polarized incident light to express E_t in terms of E_0:

$$E_x = \frac{2n_1 \cos\theta \sqrt{n_2^2 - n_1^2 \sin^2\theta}}{n_2^2 \cos\theta + n_1 \sqrt{n_2^2 - n_1^2 \sin^2\theta}} E_0 \tag{6.4a}$$

$$E_z = -\frac{2n_1^2 \sin\theta \cos\theta}{n_2^2 \cos\theta + n_1 \sqrt{n_2^2 - n_1^2 \sin^2\theta}} E_0. \tag{6.4b}$$

We can now make the transition to the supercritical regime. Note that it is not at all clear how we could make the transition to the supercritical regime by directly using Equation 6.3 since the angle of

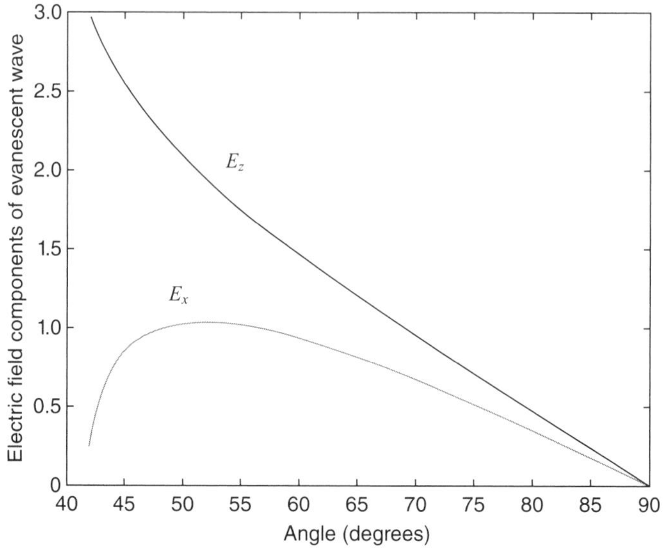

Figure 6.3 Electric field components of evanescent wave for $n_1 = 1.5$ and $n_2 = 1$.

refraction does not exist in the supercritical regime. Again, as in the case of the *s*-polarized incident light, the transition to the supercritical regime makes the square root term in Equation 6.4 imaginary—thus, the *x*- and *z*-components of the evanescent wave's electric field are out of phase with each other as well as with the electric field of the incident wave.

From Figures 6.1 and 6.3, we see that both E_z and E_y have maximum values at the critical angle and uniformly fall to zero at grazing incidence. In contrast, E_x vanishes at both the critical angle and the grazing incidence. We also see that E_z is the largest of the three components of the electric field near the critical angle.

The values of the electric field components of the evanescent wave at the critical angle are

$$E_x = 0, E_y = 2E_0, E_z = 2\frac{n_1}{n_2}E_0. \tag{6.5}$$

The result (Eq. 6.5) states that only the *z*-component of the electric field of the evanescent wave depends on the two optical materials. Thus, only the strength of the electric field of the evanescent wave perpendicular to the interface (E_z) can be influenced by the choice of optical materials. For instance, by using Ge ($n_{Ge} = 4$) as the first medium and air ($n_{Air} = 1$) as the second, we find that the *z*-component of the electric field of the evanescent wave is eight times as strong as the electric field of the incident wave.

This dramatic increase in the strength of the z-component of the electric field of the evanescent wave near the critical angle is important in spectroscopic applications of internal reflection.

Let us now look a little closer at the unusual features of the p-polarized evanescent wave. We have already seen that Maxwell equations have a solution, which is a propagating electromagnetic wave. Electric and magnetic fields of this wave are mutually perpendicular and also perpendicular to the direction of propagation. It is therefore surprising that we now find that the p-polarized evanescent wave, which in our case propagates along the x-axis, has an electric field component in the direction of propagation. This seems to be in contradiction with our previous result that an electromagnetic wave is a transversal wave.

So let us look in more detail into the electric fields of the p-polarized wave (Eq. 6.4). From Equation 6.4, it immediately follows that

$$E_x = -\frac{\sqrt{n_2^2 - n_1^2 \sin^2\theta}}{n_1 \sin\theta} E_z. \tag{6.6}$$

In the supercritical regime, the square root in Equation 6.6 is purely imaginary, implying, as we already mentioned, that E_x is oscillating 90° out of phase with E_z. Thus, when E_z takes a maximum value, E_x is zero, and vice versa. If we imagine the oscillations of the electric field play out in time, the electric field vector rotates in the plane of incidence, the tip of the vector describing an ellipse. The semiaxes of the described ellipse are given in Equation 6.4. The evanescent wave propagates along the interface oscillating between the purely transversal wave and the purely longitudinal wave. So how is it possible that the p-polarized evanescent wave seemingly violates the transverse nature of the electromagnetic wave, the result obtained directly from the Maxwell equations?

The answer is that, surprisingly, the apparent violation is not borne out when a careful calculation is done. It is true that the evanescent wave propagates only along the interface, but it does not mean that the wave vector has only the component parallel to the interface (k_x). There is also the component of the wave vector that is "perpendicular" to the interface. This component is imaginary (in mathematical sense), but it is not zero. We have also seen that the electric field component along the x-axis is also imaginary. Thus, the scalar product $\mathbf{k} \cdot \mathbf{E}$ of the wave vector and the electric field of the p-polarized evanescent wave can be evaluated by using k_x and k_z from Equation 5.2 and E_x and E_z from

Equation 6.4. The result is that the scalar product is zero. By definition, this means that the two vectors are orthogonal as required by the Maxwell equations. This somewhat unexpected turn of events emphasizes that we should not ignore the components of the wave vector that are purely imaginary. This also shows the power of the complex number-based formalism used to describe electromagnetic waves.

7 Anatomy of ATR Absorption

7.1 ATTENUATED TOTAL REFLECTION (ATR) REFLECTANCE FOR *S*- AND *P*-POLARIZED BEAM

We have seen that supercritical internal reflection is total for a nonabsorbing rarer medium. We have also seen that, in a supercritical regime, the evanescent wave is formed in the rarer medium, travels along the interface, and extends a very short distance into the rarer medium. If the rarer medium is absorbing, then the evanescent wave that travels in that medium is absorbed.

The energy flowing in an incident wave that is partially absorbed in the rarer medium is lost and is no longer present in the reflected wave. So the reflectance of the interface is no longer total. For those wavelengths for which the rarer medium absorbs, the reflectance of the interface has dips similar to those of transmittance. The total reflectance is said to have become attenuated. Actually, the absorbance derived from ATR reflectance more closely resembles the sample's absorbance index than the absorbance derived from transmittance for the following reason. In the measurement of the transmittance, what is measured is the ratio of the light power on the detector transmitted through the sample and the light power incident on the sample. But, since the sample generally reflects back some of the incident light, this reflected light does not make it to the detector and is undistinguishable from the light that was absorbed by the sample. In other words, the reflection of light from the sample makes the sample look like it absorbs light stronger than it does. Even if the sample does not absorb at all, some of the incident light is reflected back. With the supercritical internal reflection, the situation is cleaner. The reflectance of the interface is total if the sample does not absorb. Hence, in supercritical internal

Internal Reflection and ATR Spectroscopy, First Edition. Milan Milosevic.

reflection spectroscopy, if light did not make it to the detector, it is because it was absorbed by the sample.

As we have already seen, if a sample absorbs, its refractive index is complex. A complex refractive index makes supercritical internal reflectance less than total. In principle, arriving at the expression for supercritical internal reflectance for absorbing samples is easy. All one has to do is to allow the refractive index of the sample to be complex. However, although this gives the exact expression for the reflectance, it does not give any insight into the mechanism of ATR absorbance. To gain insight into the mechanism of light absorption in ATR spectroscopy, we approach the problem in three different ways. First, we analyze the exact expression for reflectance in the case of weakly absorbing samples. What we mean by weakly absorbing is that the absorption index of the sample is sufficiently small so that we can neglect the terms involving square or higher powers of absorption index. We derive the so-called weak absorption approximation keeping only the linear term in the absorption index. That gives us insight in what are the important factors governing the mechanism of absorption in supercritical internal reflection. Another approach we explore considers the absorption of the electromagnetic energy of the evanescent wave taking place on the sample side of the reflecting interface. This method enables connecting the absorbance of light by a specific element of the sample volume to the overall reflectance of the interface. The third approach further explores the leakage of electromagnetic energy through the interface indicated in Equation 5.8. The weak absorption approximation is instrumental in these analyses as the exact expression is too complicated to be illuminating.

We start the analysis by restating the expressions for supercritical internal reflectance for the *s*- and *p*-polarized light:

$$R_s = \left|r_{12}^s\right|^2 = \left|\frac{n_1\cos\theta - i\sqrt{n_1^2\sin^2\theta - n_2^2}}{n_1\cos\theta + i\sqrt{n_1^2\sin^2\theta - n_2^2}}\right|^2 \tag{7.1a}$$

$$R_p = \left|r_{12}^p\right|^2 = \left|\frac{n_2^2\cos\theta - in_1\sqrt{n_1^2\sin^2\theta - n_2^2}}{n_2^2\cos\theta + in_1\sqrt{n_1^2\sin^2\theta - n_2^2}}\right|^2. \tag{7.1b}$$

In the above expressions, the refractive index of the incident medium n_1 is assumed to be real, and the refractive index of the sample is assumed to be complex, $n_2 = n + i\kappa$. It is clear that by inserting this into the above equations, is not going to add much to our understanding of ATR.

7.2 ABSORBANCE TRANSFORM OF ATR SPECTRA

Another point has to be made on using the absorbance transform on ATR spectra. We have seen earlier that the absorbance transform is a natural transform for a transmission spectrum. It makes the resulting absorbance spectrum a linear function of both the path length through the sample and of the absorption index of the sample. This follows directly from Equations 1.11 and 1.12 and has far-reaching consequences for spectroscopy. There is no similar justification for taking the absorbance transform of ATR spectra. However, since ATR spectra in many ways resemble transmission spectra, it has become standard procedure to process ATR spectra using the same absorbance transform used for transmission spectra. Figure 7.1 shows the angular dependence of internal reflectance for *s*-polarized light. For the nonabsorbing sample, the transition between the subcritical and supercritical regimes of internal reflection is sharp. However, for even a weakly absorbing sample, there is a considerable effect on reflectance in the vicinity of the critical angle. The effect of absorption is much more pronounced on the supercritical side and is virtually undetectable on the subcritical side.

We will see below that the enhanced sensitivity of the supercritical internal reflectance on the absorption index of a sample is the

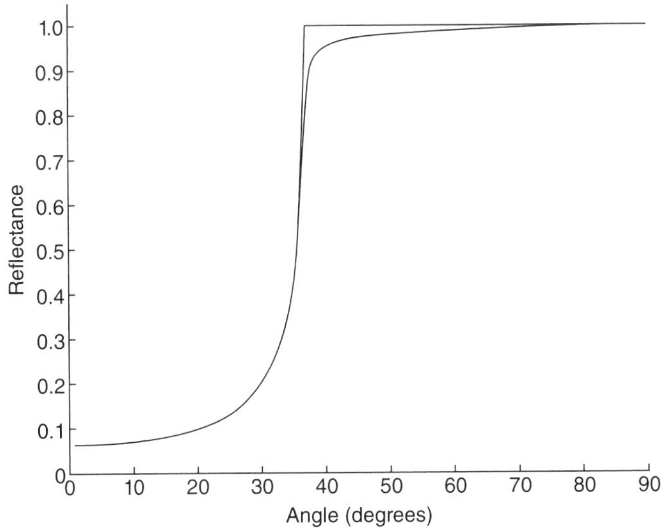

Figure 7.1 Internal reflectance of *s*-polarized light. Refractive index of incident medium $n_1 = 2.5$ and refractive index of sample $n = 1.5$ for a nonabsorbing sample ($\kappa = 0$) and for a weakly absorbing sample ($\kappa = 0.01$).

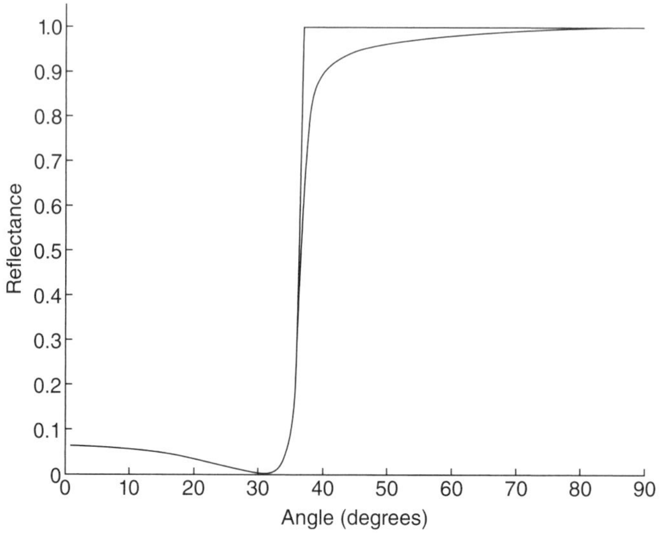

Figure 7.2 Internal reflectance of p-polarized light. Refractive index of incident medium $n_1 = 2.5$ and refractive index of sample $n = 1.5$ for a nonabsorbing sample ($\kappa = 0$) and for a weakly absorbing sample ($\kappa = 0.01$).

consequence of an unusual transition of the functional dependence of reflectance that occurs at the critical angle.

Figure 7.2 shows the internal reflectance of p-polarized light for circumstances identical to those of Figure 7.1. We see again that the effect of the absorption index is more pronounced on the supercritical than on the subcritical side. It also appears that the effect of absorbance is stronger for p- than for s-polarized light.

The vanishing of the reflectance at Brewster's angle appears unaffected by the weak absorption of the sample.

7.3 WEAK ABSORPTION APPROXIMATION

We have seen above that, even for a small value of the absorption coefficient, there is a significant effect on the supercritical internal reflection. The cumbersome looking expressions (Eq. 7.1) could be simplified for small absorbance where quadratic and higher order terms in κ can be ignored. The square root term in Equation 7.1 contains a complex number under the square root. This is because n_2 is complex and the imaginary part of the expression under the square root is proportional to κ. Since we are assuming that κ is small, following Equations 5.7 and 5.9 to the first order in κ, we can write

$$\sqrt{n_1^2\sin^2\theta-n_2^2}=\sqrt{n_1^2\sin^2\theta-n^2}-\frac{in\kappa}{\sqrt{n_1^2\sin^2\theta-n^2}}. \tag{7.2}$$

Using Equation 7.2, Equation 7.1 can then be simplified to

$$R_s=1-\frac{4nn_1\kappa\cos\theta}{(n_1^2-n^2)\sqrt{n_1^2\sin^2\theta-n^2}} \tag{7.3a}$$

$$R_p=1-\frac{4nn_1\kappa(2n_1^2\sin^2\theta-n^2)\cos\theta}{(n_1^2-n^2)[(n_1^2-n^2)\sin^2\theta-n^2]\sqrt{n_1^2\sin^2\theta-n^2}}. \tag{7.3b}$$

Expressions (Eq. 7.3) for supercritical internal reflectance are correct to linear terms in the absorption index κ. We can now examine how various parameters such as the refractive index of the internally reflecting medium n_1, the refractive index of the sample n, the angle of incidence and polarization of incident light control the reflectance. The second term on the right-hand side of the expressions (Eq. 7.3) is the one that accounts for absorption of light in ATR. This term is linear in the absorption index of the sample. This may sound like tautology since we derived the above expressions by keeping only the linear term; however, it could have happened that the linear term was zero, so the fact that the linear term is nonvanishing is important. Note that the absorbance transforms of the above expressions, again to linear terms in κ, are

$$A_s=0.434\frac{4nn_1\kappa\cos\theta}{(n_1^2-n^2)\sqrt{n_1^2\sin^2\theta-n^2}} \tag{7.4a}$$

$$A_p=0.434\frac{4nn_1\kappa(2n_1^2\sin^2\theta-n^2)\cos\theta}{(n_1^2-n^2)[(n_1^2-n^2)\sin^2\theta-n^2]\sqrt{n_1^2\sin^2\theta-n^2}}, \tag{7.4b}$$

where we used the approximation valid for small x:

$$-\log(1-x)\cong 2.303x.$$

Equations 7.3 and 7.4 are of central importance for ATR spectroscopy.

The linear relationship between the absorbance and the absorption index shown in Equation 7.4 is approximate and only valid in the weak absorption approximation. The validity range of the weak absorption approximation can be assessed by comparing the approximate expressions (Eq. 7.4) with the exact expressions (Eq. 7.1). Figure 7.3 shows

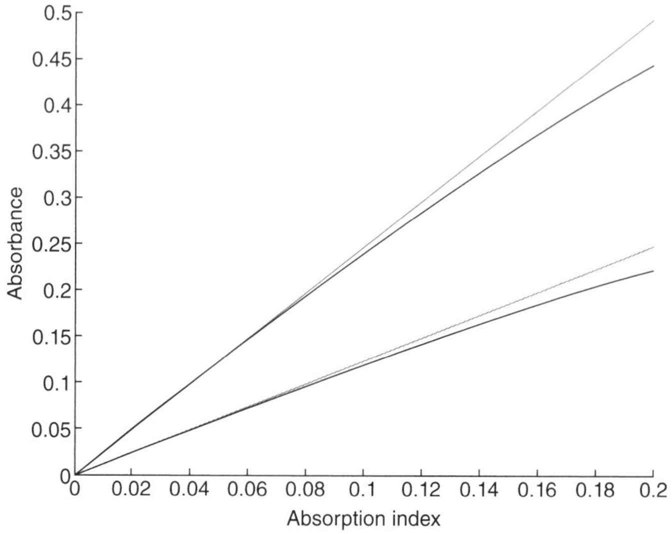

Figure 7.3 Comparison between absorbances determined by the weak absorption approximation (straight lines) and by the exact expressions for two polarizations. The angle of incidence is 45°; the refractive index of ATR material is 2.5; and the real part of the refractive index of sample is 1.5.

this comparison for s and p polarization. The exact expressions exhibit slight nonlinearity. The graphs of weak approximation expressions are straight lines tangential to the graphs of the exact curves. The lower curve is the graph of the expression for the s-polarized light. The absorbance for the p-polarized light appears twice as strong as the absorbance for s-polarized light. Actually, it turns out that for the angle of incidence of 45°, the absorbance for p-polarized light is exactly twice of that for s-polarized light. This follows directly from Equation 7.4. By inserting $\theta = 45°$, the ratio of two expressions unexpectedly simplifies to

$$\frac{A_p}{A_s} = \frac{2n_1^2 \sin^2\theta - n^2}{\left(n_1^2 + n^2\right)\sin^2\theta - n^2} = 2. \tag{7.5}$$

We have encountered this unusual relationship between the reflectance for s and p polarizations earlier, when we found that, for the angle of incidence of 45°, the reflectance for the p-polarized light is the square of the reflectance for the s-polarized light. The absorbance transform of reflectance is essentially equivalent to taking the logarithm of reflectance, which explains the factor of 2 found above for both the exact expressions and for the weak absorption limit.

7.4 SUPERCRITICAL REFLECTANCE AND ABSORPTION OF EVANESCENT WAVE

Although it was straightforward to derive the weak absorption results (Eqs. 7.3 and 7.4), we learned little about the mechanism of absorption that takes place in the regime of supercritical internal reflection. It is clear that the attenuation of the total reflectance is due to the interaction of the evanescent wave and the absorbing medium. However, we would like to gain a more detailed understanding of the absorption process and where and how it occurs. Figure 5.3 illustrates the basic geometry of the internal reflection. A beam of cross section D is incident on the reflecting interface at a supercritical angle, θ. The area of the interface illuminated by the incident beam is $D/\cos\theta$. The illuminated area of the interface supports the evanescent beam on the other side of the interface. The energy density associated with the evanescent wave,

$$u(z)=\frac{\varepsilon_2}{4\pi}|E_t(z)|^2=\frac{n_2^2}{4\pi}|E_t(0)|^2 e^{-\frac{2z}{d_p}} \\ =\frac{n_2^2}{4\pi}|t_{12}|^2|E_0|^2 e^{-\frac{2z}{d_p}} \tag{7.6}$$

falls off exponentially with distance z from the interface. As we have seen in Equation 3.10, the rate at which the energy is absorbed by the sample is proportional to the energy density itself. If we integrate this rate of energy absorption over the volume filled by the evanescent wave, we get the energy absorbed by the sample per unit time, or the absorbed power. The power absorbed by the sample is thus absent from the reflected wave. The absorbed power is

$$P_A=\int\frac{du}{dt}dV=-4\pi\kappa k\frac{c}{n}\int u dV=-\frac{Dn\kappa kc}{\cos\theta}|t_{12}|^2|E_0|^2\int_0^\infty e^{-\frac{2z}{d_p}}dz \\ =-\frac{Dn\kappa kc}{\cos\theta}|t_{12}|^2|E_0|^2\frac{d_p}{2}. \tag{7.7}$$

The incident power is

$$P_{in}=\frac{|E_0|^2}{4\pi}cn_1D. \tag{7.8}$$

So, the fraction of incident power absorbed by the sample is

$$\frac{P_A}{P_{in}} = \frac{n\kappa|t_{12}|^2}{n_1\cos\theta\sqrt{n_1^2\sin^2\theta - n^2}}. \tag{7.9}$$

For s-polarized incident light, the straight substitution of t_{12}^s from Equation 4.8 yields

$$|t_{12}^s|^2 = \frac{4n_1^2\cos^2\theta}{n_1^2 - n^2}. \tag{7.10}$$

The expression (Eq. 7.10) inserted into Equation 7.9 leads immediately to Equation 7.4a. The factor 0.434 in Eq. 7.4 is the consequence of using the decadic rather than the natural logarithm in the definition of absorbance.

However, the same procedure applied to p-polarized light does not lead to the correct expression. From the expression (Eq. 4.10) for t_{12}^p, we get

$$|t_{12}^p|^2 = \frac{4n_1^2 n_2^2\cos^2\theta}{(n_1^2 - n^2)[(n_1^2 + n^2)\sin^2\theta - n^2]}, \tag{7.11}$$

which does not have the correct numerator. This is a puzzling result that requires us to reexamine the derivation that resulted in Equation 7.11.

If we use the expressions (Eq. 6.4) for the electric field components E_x and E_z, we find

$$|t_{12}^p|^2 = \frac{|E_x|^2 + |E_z|^2}{|E_{in}|^2} = \frac{4n_1^2(2n_1^2\sin^2\theta - n^2)\cos^2\theta}{(n_1^2 - n^2)[(n_1^2 + n^2)\sin^2\theta - n^2]}. \tag{7.12}$$

The two expressions, Equations 7.11 and 7.12, are different, and what adds to the puzzle is that we actually derived the expressions (Eq. 6.4) for E_x and E_z by utilizing the expression (Eq. 4.10) for t_{12}^p to find E_t, which we then projected onto the x- and z-axes to find E_x and E_z, respectively. So it is puzzling that by using the field components, we arrive at the correct expression, while by using the field itself, we arrive at the incorrect result.

The answer to the puzzle is that the components were projected using the angle of refraction φ and that this angle loses its meaning in the supercritical regime. So, if vector E_t^p is expressed in terms of its components E_x and E_z as done in Equation 6.3, the result is

$$|E_t^p|^2 = |E_x|^2 + |E_z|^2 = |E_t^p|^2(\cos^2\varphi + \sin^2\varphi). \tag{7.13}$$

The expression (Eq. 7.13) looks like a simpleminded exercise since

$$\cos^2\varphi + \sin^2\varphi = 1,$$

but notice what happens if we reexpress the angle φ in terms of the angle of incidence θ using Snell's law:

$$\begin{aligned}\cos^2\varphi + \sin^2\varphi &= \left|\frac{\sqrt{n^2 - n_1^2\sin^2\theta}}{n}\right|^2 + \frac{n_1^2\sin^2\theta}{n^2} \\ &= \left|\frac{n^2 - n_1^2\sin^2\theta}{n^2}\right| + \frac{n_1^2\sin^2\theta}{n^2}.\end{aligned} \tag{7.14}$$

The expression (Eq. 7.14) can be evaluated for two distinct cases. In the subcritical regime of internal reflection, where $n > n_1\sin\theta$, the expression under the absolute value is positive and the absolute value sign can be dropped, leading to the standard result of one. However, in the supercritical regime, $n < n_1\sin\theta$ so the right-hand side of Equation 7.14 now takes form

$$\frac{n_1^2\sin^2\theta - n^2}{n^2} + \frac{n_1^2\sin^2\theta}{n^2} = \frac{2n_1^2\sin^2\theta - n^2}{n^2}. \tag{7.15}$$

This unexpected result is the consequence of the angle of refraction not being a real number but imaginary, so the usual relationship $\cos^2\varphi + \sin^2\varphi = 1$ breaks down and is no longer valid. This confirms that the expression (Eq. 7.12) is indeed the correct result for the supercritical case of internal reflection, while Equation 7.11 remains valid in the subcritical regime.

Now it is straightforward to insert Equation 7.12 into Equation 7.9 to arrive at Equation 7.4b. So, with the above clarification, we have successfully demonstrated that these two different calculations of the fraction of absorbed power lead to the same result. That justifies our intuitive sense that in the phenomenon of supercritical internal reflection, the reduction in the reflected intensity is due to the absorption of the evanescent wave by the sample.

We calculated Equations 7.3 and 7.4 from the exact results (Eq. 7.1) by applying the weak absorption approximation and by keeping only the terms linear in the absorption index. The calculation based on the absorption of the evanescent wave seemed not to have been based on the weak absorption approximation. However, by reexamining carefully the above derivation, we see that for the evanescent wave, we used the result valid for a nonabsorbing rarer medium, which in effect constitutes the weak absorption approximation. We have also learned that the absorption of incoming light intensity is proportional to the intensity of the evanescent wave, which is highest at the very interface and decays exponentially with distance into the absorbing sample. The exponential decay of the intensity of the evanescent wave is faster than the decay of the electric field. The penetration depth for the electric field of the evanescent wave is d_p, while the penetration depth for the intensity is $d_p/2$ implying that ~63% of light is absorbed within the sample layer $d_p/2$ thick. When it comes to ATR spectroscopy, it would be more appropriate to define penetration depth as half of the conventional value (Eq. 5.4). However, to avoid confusion, we will adhere to the conventional definition of penetration depth.

7.5 THE LEAKY INTERFACE MODEL

We now turn to the third way to analyze absorption occurring in supercritical internal reflection. This time, we analyze the phenomenon by observing how much of the incident beam is "leaking" through the interface. The expression for the electric field of the evanescent wave (Eq. 5.6), in the case when the sample in contact with the ATR material is absorbing, contains a factor describing the flow of a component of the evanescent wave that is not parallel but is perpendicular to the interface. We saw that this component is proportional to the absorption index of the sample. This flow represents energy leaking through the interface to become absorbed by the sample. Clearly, if there is absorption occurring in the sample on the other side of the interface, this absorbed power has to be replenished by additional energy flowing through the interface. The electromagnetic power that flows through the interface thus must equal the power absorbed by the sample. And this is true for any plane in the sample that is parallel to the interface; that is, the flow of power through such a plane must be equal to the power absorbed by the sample beyond that plane.

To analyze the power that "leaks" through the interface and into the sample, we start with the incident beam. The power carried by the incident beam of cross-sectional area D illuminates the reflecting area of the interface $D/\cos\theta$. The power density of the incident wave is given by Equation 4.13. When the angle of incidence θ crosses the threshold for supercritical internal reflection, the transmitted wave collapses into the evanescent wave. The electric field of the evanescent wave is connected to the electric field of the incident wave by the Fresnel amplitude coefficient t_{12}. If a sample is nonabsorbing, the flow of energy in the evanescent wave is parallel to the interface, so no transmission of energy through the interface takes place as can be seen from the expression (Eq. 4.16). The square root term in Equation 4.16 is imaginary for the supercritical incidence. It is not clear what the meaning of the imaginary transmittance could be since transmittance is defined as the ratio of two absolute values and, by definition, has to be a real positive number smaller or equal to one. A standard way to handle this is to retain only the real part of the expression. If the expression for transmittance is purely imaginary, then the real part is zero. However, if the sample absorbs, the square root term in Equation 4.16 is no longer purely imaginary, it become complex having both real and imaginary parts as shown in Equation 5.7. The real part of the expression for transmittance (Eq. 4.16) is then

$$T = \frac{n\kappa |t_{12}|^2}{n_1 \cos\theta \sqrt{n_1^2 \sin^2\theta - n^2}}, \tag{7.16}$$

where we have used the result (Eq. 5.9) valid for small κ to find the real part of the square root term. The expression (Eq. 7.16) gives the weak absorption approximation for the fraction of the incident power that leaks through the interface and into the sample. Since the sample absorbs light, any light that leaks through the interface is completely absorbed by the sample. By going back to the expression (Eq. 7.9) that gives the fraction of incident power that is absorbed by the sample, we see that Equation 7.16 is identical to Equation 7.9. Equation 7.9 was the starting point for deriving the expressions (Eq. 7.4) for absorbed power, so we do not have to repeat the derivation. The conclusion is that we have shown that the expressions (Eq. 7.4) for absorbed power in ATR can be arrived at by evaluating the fraction of radiative power that leaks through the interface. This leakage is associated with the fact that the evanescent wave in an absorbing medium acquires a component that propagates perpendicular to the interface, as we have found in Equation 5.8. Inside the sample, at the distance z from the interface, this flow of energy is

$$P_z(z) = \frac{|t_{12}|^2 e^{-\frac{2z}{d_p}}}{n_1 \cos\theta} Re\left(i\sqrt{n_1^2 \sin^2\theta - n^2}\right) P_{in}. \tag{7.17}$$

Thus, as the radiative power flows into the sample, it is reduced by the exponential term. This is in agreement with what we have found in Equation 7.7, where we were evaluating the total power absorbed by the sample. In that case, we were integrating the energy density of the

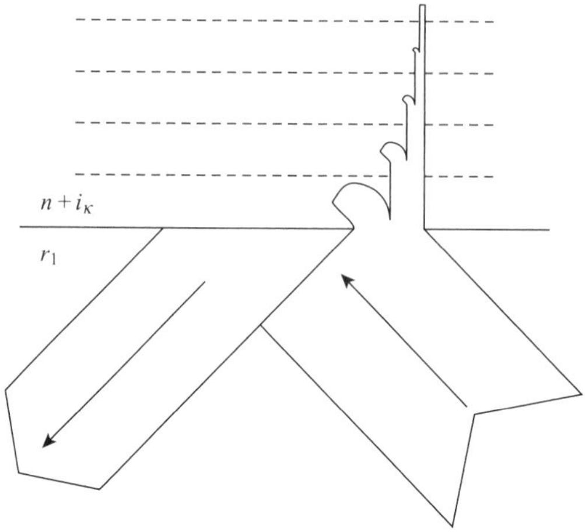

Figure 7.4 Schematic representation of the energy flow in ATR.

evanescent wave over the volume containing the wave. Since energy density in the evanescent wave falls off as

$$e^{-\frac{2z}{d_p}}$$

with the distance from the interface, so does the rate at which the radiative power is absorbed by the sample. Thus, it would be expected that a new energy has to flow into the volume of the sample at the same rate at which the energy is being absorbed. Figure 7.4 shows schematically how the energy from the incoming light beam flows into the sample where it is absorbed at different rates at different depths. The flow of energy into the sample perpendicular to the interface supplies the lost energy into the evanescent wave flowing along the interface. The reflected light beam carries less power than the incoming beam, the difference being absorbed by the sample.

We have now examined three different ways of analyzing supercritical internal reflection at the interface of an ATR material with an absorbing sample. We have shown that all three approaches lead to the same result for reflectance. We have also seen that absorption takes place in the sample very near to the interface. In the process of energy absorption, the power is drained out of the evanescent wave. This power is replenished by the leakage of the power of the incident beam through the interface. This leakage of light energy through the interface is the reason why the interface is not totally reflective.

8 Effective Thickness

8.1 DEFINITION AND EXPRESSIONS FOR EFFECTIVE THICKNESS

We have seen that reflectance in attenuated total reflection (ATR) in many ways resembles transmittance in transmission spectroscopy. In a transmission experiment, if light is absorbed by a sample, then the light reaching the detector is weaker than the light reaching the detector in the absence of the sample. If the sample absorbs at a particular wavelength, the transmittance of the sample exhibits a downward-looking peak in the spectrum at that wavelength.

Similarly, the same sample analyzed by ATR will exhibit a downward-looking peak in the reflectance spectrum at or close to the same wavelength. Thus, it would be useful to evaluate how the two absorbances for this peak, one measured in transmission and one in ATR, are mutually related. In other words, if we measure a spectrum of a sample in ATR and convert it into absorbance and find the peak to be of certain strength, what would the sample thickness have to be so that peak of the same strength is measured in transmission. The transmittance of a sample of thickness d is given by Equation 1.11 and thus the absorbance is given by Equation 1.12. If the same sample is measured by ATR, the situation is a bit more complex since the reflectance measured depends on the angle of incidence and the polarization of the incident light and also on the refractive index of the ATR element. Let us consider the case of weak absorption so we could use expressions (Eq. 7.4) for absorbance as measured in ATR. For the absorbances measured in the two cases to be equal, Equation 1.12 must be equal to Equation 7.4. For the s-polarized beam, that means

$$\pi k d_{\text{eff}}^{s} = \frac{n n_1 \cos\theta}{\left(n_1^2 - n^2\right)\sqrt{n_1^2 \sin^2\theta - n^2}}, \tag{8.1}$$

where we have removed the factors that are the same on both sides and we identify this specific thickness as the effective thickness d_{eff}^{s} for the s-polarized light. Similarly, for the p-polarized light, we find

$$\pi k d_{\text{eff}}^{p} = \frac{n n_1 \left(2n_1^2 \sin^2\theta - n^2\right)\cos\theta}{\left(n_1^2 - n^2\right)\left[\left(n_1^2 + n^2\right)\sin^2\theta - n^2\right]\sqrt{n_1^2 \sin^2\theta - n^2}}. \tag{8.2}$$

The concept of effective thicknesses is different from the concept of penetration depth d_p. The penetration depth was defined as a measure of the quickness of decay of the strength of the electric field of the evanescent wave with the distance from the interface into the sample. This depth has the clearest meaning for a nonabsorbing sample. The presence of absorption simply complicates the situation as the expression becomes complex when the refractive index of the sample acquires a nonvanishing imaginary part. In the case of an absorbing sample, we can redefine the expression (Eq. 5.4) to specify that only the real part of the expression be used, or that we use the absolute value of the expression so that it remains meaningful. On the other hand, if the sample is nonabsorbing, the identification as we did in Equations 8.1 and 8.2 would not be meaningful. Both sides of the equation would be zero. So, the definition of effective thickness only makes sense if the sample is absorbing.

8.2 EFFECTIVE THICKNESS AND PENETRATION DEPTH

It is hard not to notice that the expressions for effective thickness as could be extracted from Equations 8.1 and 8.2 clearly contain the expression for the penetration depth; that is,

$$d_{\text{eff}}^{s} = \frac{2 n n_1 \cos\theta}{\left(n_1^2 - n^2\right)} d_p \tag{8.3}$$

and

$$d_{\text{eff}}^{s} = \frac{2 n n_1 \left(2n_1^2 \sin^2\theta - n^2\right)\cos\theta}{\left(n_1^2 - n^2\right)\left[\left(n_1^2 + n^2\right)\sin^2\theta - n^2\right]} d_p. \tag{8.4}$$

The expressions for effective thickness, Equations 8.3 and 8.4, summarize the important features of ATR. Even though these expressions are approximations and are strictly valid only for weak absorptions, the conclusions derived from them are valid generally.

Another variation of the expression for the effective thickness provides a further insight into the relevant terms:

$$d_{\text{eff}}^{\text{s,p}} = \frac{n_1}{n \cdot \cos\theta} \left| t_{12}^{\text{s,p}} \right|^2 \frac{d_p}{2}, \tag{8.5}$$

where we combined the two expressions for different polarizations into one. The individual expressions for the *s* and *p* polarizations are obtained by choosing either *s* or *p* in the superscripts of d_{eff} and t_{12}. Thus, the effective thickness is proportional to the penetration depth and to the absolute value squared of the transmission amplitude coefficient, the factor that measures the strength of coupling of the incident wave to the evanescent wave. The refractive indices account for changes in energy density in going from one material to another. In addition, there is the $1/\cos\theta$ factor that accounts for the change of the area illuminated by the incident beam as a function of the angle of incidence. This factor affects the volume in which the evanescent wave exists and thus the total absorbed power. This factor would make the effective thicknesses infinite at the grazing incidence were it not for the counteracting term $\cos^2\theta$ contained in $\left|t_{12}^{s,p}\right|^2$. So, in the end, the effective thicknesses become zero at the grazing incidence for both polarizations. The penetration depth, on the other hand, controls the behavior of the effective thicknesses near the critical angle, making them both infinite at the critical angle. The critical angle itself depends on the refractive indices of the sample and the ATR crystal. This dependence on the two refractive indices n_1 and n can be reduced to the dependence on the ratio n/n_1. For all practical purposes, one could assume that the refractive index of the sample is $n = 1.5$. The refractive index of the ATR material boils down to several practical choices. Since the ATR material has to have a high refractive index and also has to be transparent in the mid-infrared (IR) spectral range, possible materials include germanium ($n_{\text{Ge}} = 4$), diamond ($n_{\text{diamond}} = 2.4$), zinc selenide ($n_{\text{ZnSe}} = 2.4$), and silicon ($n_{\text{Si}} = 3.45$). There are really only a handful of practically available ATR materials. Figure 8.1 shows the effective thickness for two common ATR materials, Ge and ZnSe, versus the angle of incidence.

Note that the effective thickness for *p* polarization is always larger than the effective thickness for *s* polarization at the same angle of incidence. Also, the lower the refractive index of the ATR material, the larger the effective thickness. The effective thickness is almost always smaller than the wavelength of light.

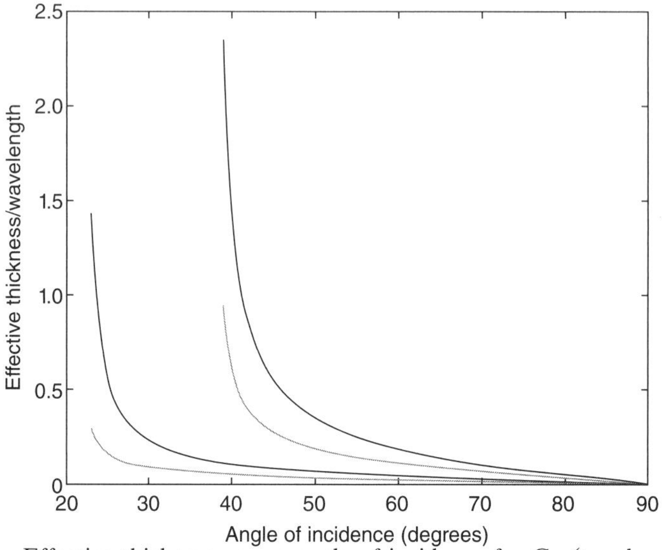

Figure 8.1 Effective thickness versus angle of incidence for Ge (two lower curves) and ZnSe ATR crystals for a typical sample with refractive index $n = 1.5$. The upper curve in each group is for *p* polarization.

8.3 EFFECTIVE THICKNESS AND ATR SPECTROSCOPY

The diverging behavior of effective thickness near the critical angle is evident for both materials shown in Figure 8.1 and is a general characteristic of ATR spectroscopy. It seems that a good strategy to increase the effective thickness in the ATR measurement would be to select an angle of incidence very close to the critical angle. However, note the steep slope in the graph of effective thickness near the critical angle. Any variation in the angle of incidence causes a significant change in the effective thickness, and there is a possibility that some rays in the incident beam may fall below the critical angle. Those rays would be in the subcritical regime of internal reflection and would transmit through the interface and into the sample, causing a dip in the reflectance that is not caused by the absorption by the sample. In the spectral regions where the sample does not absorb, this causes a baseline shift in the absorbance spectrum of the sample. Since the refractive indices of both the ATR element and the sample generally change with the wavelength, the fraction of the rays falling below the critical angle will also change with the wavelength, causing the baseline shift to change with the wavelength. However, this change is slow and easy to remove from the spectra. The problem is the change in the refractive index that occurs around the absorption band. The situation is shown in Figure

2.3, where the change in the real part of the refractive index in the vicinity of the absorption band is illustrated. On the left side of the peak, the refractive index dips below the level of the refractive index outside the absorption region, while on the right side, it peaks above that level. Thus, even if all the rays in the incoming beam are safely above the critical angle for the areas of the spectrum where the sample is not absorbing, the situation changes in the vicinity of the absorption band. On the right side of the peak, the refractive index of the sample sharply increases, thereby increasing the critical angle for those wavelengths. The rays at those wavelengths could easily fall below the critical angle and thus transmit into the sample. This would make the apparent reflectance observed for these wavelengths very low. Even if these rays remain above the critical angle, the effective path length for these rays is larger than the effective path length calculated for the center of the peak, making the right side of the absorption peak heavier. On the left side of the peak, the refractive index dips, making the effective path length on the left side of the peak smaller than for the center of the peak. The net result is that the absorption peak becomes distorted and the apparent peak position shifts to lower wave numbers. The closer the angle of incidence is to the critical angle, the larger the distortion of the band shape and the greater the shift of the apparent peak maximum to the lower wave numbers.

The important feature of ATR spectroscopy, as we have seen, is that the effective thickness associated with ATR is a fraction of a wavelength. It is possible to produce subwavelength transmission path lengths for liquids by placing an ultrathin spacer between two transparent windows. However, getting a liquid sample into such a narrow gap is usually not easy especially if the sample is viscous. And even if one succeeds getting the sample between the two windows, the effect of a thin sample in a beam is to produce so-called interference fringes. These are an undulating pattern superimposed on the baseline of the spectrum. We will further investigate this effect later on. Another problem such a short path length transmission setup would encounter is that it would be difficult to force a sample to flow through such a narrow gap. The flow rate through such a small cross section would be extremely low and would require high pressure to sustain it. Many samples, such as water, absorb strongly in the mid-IR spectral region, and path lengths as short as 10 or 25 μm are required for their analysis. The fact that ATR spectroscopy provides such short path lengths naturally and reproducibly is one of the principal reasons for the popularity of ATR spectroscopy. The sample is simply placed on the ATR crystal, and as long as the sample thickness exceeds a few times the penetration depth, the precise thickness of the sample is irrelevant. The sample can be liquid or gel or it can even be solid.

8.4 EFFECTIVE THICKNESS FOR STRONG ABSORPTIONS

The effective thickness exists only for angles of incidence above the critical and is defined only in the low absorption approximation. The definition of effective thickness could be extended to stronger absorptions as follows:

$$d_{\text{eff}}^{s,p} = \frac{-\log\left(R^{s,p}\right)}{4\pi k\kappa}.$$

Note that this expanded definition reduces to the usual low absorption expression in the case of weak absorptions. A consequence of this expanded definition of effective thickness would be that it now becomes dependent on the sample absorption index and hence is not particularly useful. The definition of effective thickness, on the other hand, cannot be extended below the critical angle. The reason is that below the critical angle, there is a portion of light that is transmitted through the interface, and the absence of energy in the reflected wave is not only due to absorption by the sample but is also carried away by the transmitted wave. Even the concept of a critical angle in the case of absorbing samples becomes fuzzy.

9 Internal Reflectance near Critical Angle

9.1 TRANSITION FROM SUBCRITICAL TO SUPERCRITICAL REFLECTION

We defined the critical angle for internal reflection in purely geometric terms as the angle for which light incident from a medium of higher refractive index onto the interface with a medium of lower refractive index refracts at the angle of 90°. This is the highest angle that light can still refract at. Above this angle, light is totally internally reflected. This sharp transition is seen in the reflectance of the interface as shown in Figures 7.1 and 7.2 for the case of a nonabsorbing sample.

Figures 7.1 and 7.2 also show the change in the behavior of reflectance for the case of an absorbing sample. The sharp transition that demarcates the subcritical from the supercritical regimes of internal reflection for a nonabsorbing sample is no longer sharp for an absorbing sample. The transition becomes gradual and it is no longer clear where the critical angle should be. However, the transition between subcritical and supercritical internal reflections remains quick and fairly dramatic. Figure 9.1 shows the optical constants of a hypothetical sample. The sample has a single absorption band in the spectral region of interest, a typical refractive index of about 1.5, and a relatively weak absorption peak. Thus, its optical constants are quite ordinary. For an illustration of the transition from subcritical to supercritical internal reflection, we show in Figure 9.2 a series of reflectance curves for angles of incidence increasing in steps of 0.1° from subcritical to supercritical reflection using the optical constants of Figure 9.1.

The optical constants of the model material are chosen to represent a typical material. The spectra shown in Figure 9.2 are internal reflection spectra for p-polarized light for our hypothetical material in contact with the internal reflection material with a refractive index of 2.42. The

Internal Reflection and ATR Spectroscopy, First Edition. Milan Milosevic.

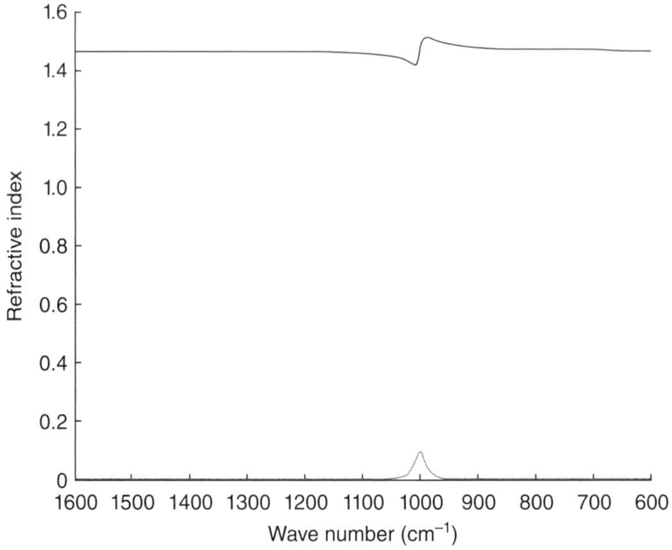

Figure 9.1 Optical constants of a model material used in simulations.

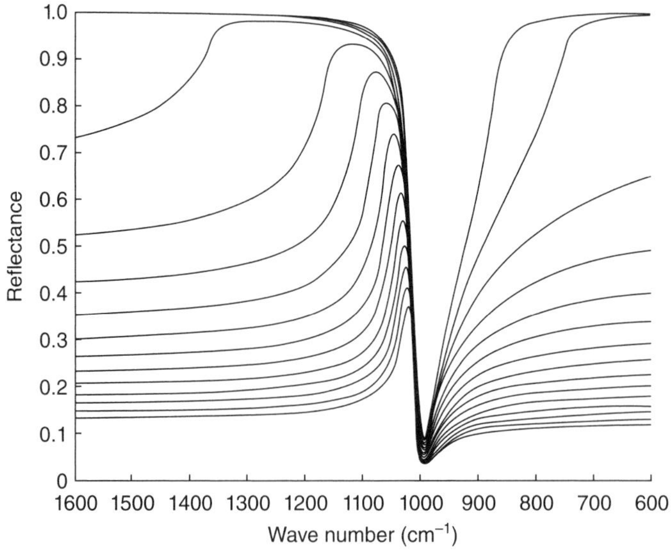

Figure 9.2 Internal reflection spectra near critical angle in 0.1° steps starting with 36.1° (lowest spectrum) and ending with 37.6° (highest spectrum).

lowest spectrum is for the angle of 36.1° and the highest spectrum is for the angle of 37.6°.

The lowest spectrum clearly looks like a subcritical internal reflection spectrum with the derivative shaped peak. The highest spectrum clearly looks like a typical supercritical internal reflection (attenuated total reflection [ATR]) spectrum with a downward pointing peak, dipping from the 100% reflectance baseline, indicating that light is totally internally reflected in the spectral regions where the sample does not absorb. The band is still somewhat distorted on the low wave number side reflecting the uncomfortable closeness of the angle of incidence to the critical angle. Actually, one could argue that the critical angle in this particular case is somewhere between 37.4° and 37.5° since the reflectance for 37.4° clearly does not return to total away from the peak, while at 37.5° it clearly does. Also, note that the absorbance associated with the internal reflection just above the critical angle is not particularly high despite the angle of incidence being so close to the critical angle.

Another important feature of the transition of the internal reflectance as a function of the angle of incidence from subcritical to supercritical reflection, as illustrated in Figure 9.2, is that reflectance increases with the increasing angle of incidence at every wave number regardless of the nature of the absorption peak. The absorption peak apparently evolves from the dip in the subcritical reflection spectrum that occurs on the low wave number side of the absorption peak. This clearly shows that the peak position in ATR is always somewhat shifted toward the lower wave numbers with respect to the true peak position. This apparent peak shift decreases as the angle of incidence increases further away from the critical, and it can be seen as a part of a broader peak distortion that in ATR occurs near the critical angle.

9.2 EFFECTIVE THICKNESS AND REFRACTIVE INDEX OF SAMPLE

Another way to view this peak distortion is to say that the effective thickness is not constant for the whole spectral region but changes through the absorption peak as the refractive index of the sample swings first down then up as the wave number sweeps through the peak position. On the high wave number side, the refractive index swings below its baseline value making the effective thickness smaller than the nominal effective thickness calculated with the refractive index away from the absorption band. On the low wave number side, the refractive index of the sample shoots above its baseline and the effective thickness grows larger than nominal.

Thus, the absorbance on the low wave number side is smaller than the absorbance on the high wave number side. If the angle of incidence is very close to critical, the increase of the refractive index of the sample on the low wave number side could cause the angle of incidence to dip below critical. This would make the reflectance at those wave numbers subcritical. Light would transmit through the interface and continue to propagate into, and eventually become absorbed by, the sample. This lost energy is absent in reflected light and appears as artificially boosted absorbance. It is true that the energy transmitted through the interface is eventually absorbed in the sample, given the sufficient sample thickness, but this light is absorbed through a different mechanism and this absorbance is not coadditive with ATR absorbance. However, the spectrometer has no way of discriminating between the two mechanisms so, generally, it is best to use the angle of incidence that is sufficiently high to keep it above critical for all the wavelengths in the spectral range of interest.

9.3 CRITICAL ANGLE AND REFRACTIVE INDEX OF SAMPLE

This situation that the reflection at some wavelengths could be subcritical, while for other wavelengths, it could be supercritical can occur in practical measurements. To investigate this a little further, we selected for our hypothetical sample somewhat more realistic optical constants shown in Figure 9.3. Note that the "wiggles" in the refractive index are not particularly strong keeping the refractive index values well between 1.40 and 1.60. Since the refractive index of the sample changes with the wave number, the critical angle changes correspondingly. The critical angle as a function of the wave number can be calculated from

$$\sin\theta_c = \frac{\sqrt{n^2 - \kappa^2}}{n_1},$$

where both n and κ are functions of the wave number. Using the above expression, the critical angle for the sample with optical constants shown in Figure 9.3 can be easily calculated and displayed as a function of the wave number. This is done in Figure 9.4. We can clearly see that even a moderate variation in the refractive index of the sample introduces several degrees of variations in the critical angle. Thus,

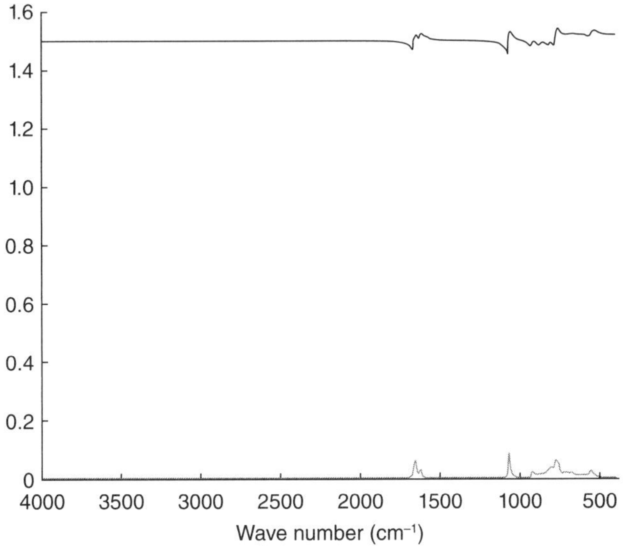

Figure 9.3 Optical constants of a hypothetical sample used in spectral simulations.

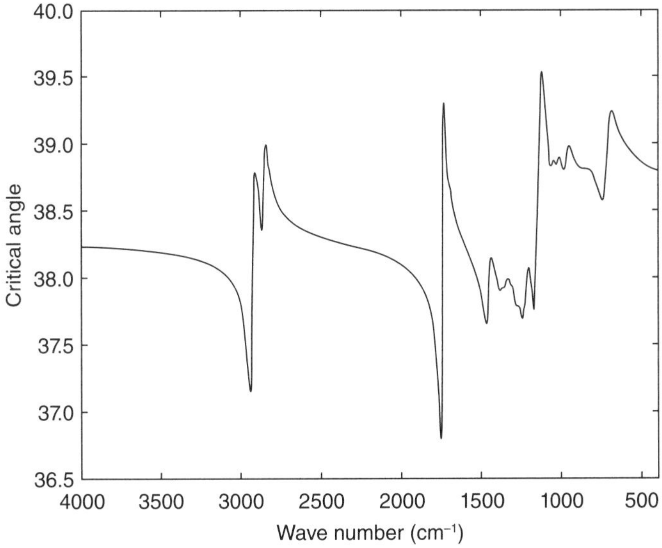

Figure 9.4 Critical angle variation for the sample from Figure 9.3.

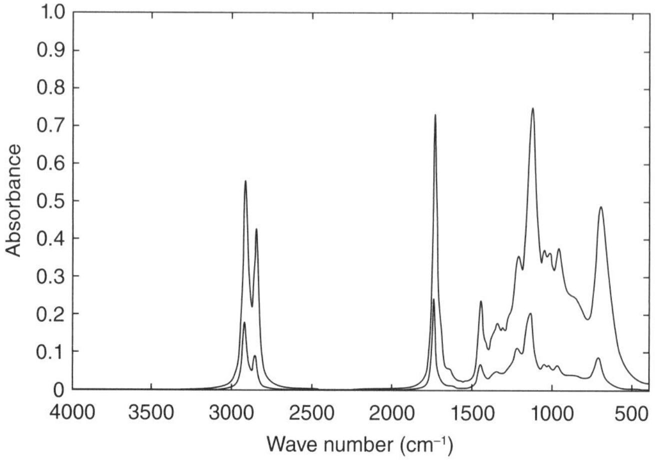

Figure 9.5 Spectra of the sample with the optical constant from Figure 9.3 for angles of incidence 39° (higher) and 45°.

assuming that all the incoming rays are incident at the same angle, for the angle of incidence above 40°, the reflection would be safely supercritical for the entire spectral range shown. However, if we selected the angle of incidence of 39°, we see that the incidence will be subcritical at a number of areas of the spectral region where the critical angle has sharp upward spikes. The question of interest is whether this crossing of the critical angle threshold will have any clearly discernable effect on the spectra. The crossing from supercritical to subcritical reflection could introduce a nonlinear response and could adversely affect any quantitative measurement. The spectra for the angle of incidence of 45° (thus safely supercritical) and 39° (piecewise subcritical) are shown in Figure 9.5. We immediately see a great difference (a factor of 3–5X) in the measured absorbance, but spectral distortions are not apparent. The increasing absorbance enhancement toward lower wave numbers is a consequence of the average critical angle approaching the angle of incidence of 39°. However, the meaning of reflectance for those spectral regions where the angle of incidence drops below the critical is not clear. There are no apparent spectral distortion in these areas; thus, no unambiguous signal that the reflectance in these spectral regions is not true ATR. We know that the quasi linearity of ATR may not extend to subcritical incidence. The question is whether the quasi linearity that is valid in the supercritical regime extends to a subcritical incidence. The answer comes from the fact that the spectrum at 39° is not distorted. If it was, it would be a clear sign that the quasi linearity does not extend to a subcritical regime.

Obviously, if we pushed further into the subcritical regime, eventually, the spectra would become distorted and the quasi linearity would be clearly disrupted. However, it appears that we can push somewhat into the subcritical regime without being punished by the breakdown of the ATR technique. Since, by pushing the angle of incidence into the subcritical regime, we benefit from a substantially increased absorbance, the question naturally arises: How far can we push without being penalized? The answer is that if the incident angle drops below critical in a narrow angular range but returns above critical on both sides of this range, we are safe and the ATR technique is operational. If not, the reflection is no longer supercritical and we lose the quasi linearity we need for quantification.

10 Depth Profiling

10.1 ENERGY ABSORPTION AT DIFFERENT DEPTHS

The detailed examination of the absorption mechanism in attenuated total reflection (ATR) that we undertook in the previous chapter has paved the way for the analysis of depth profiling by ATR. We analyzed how the evanescent wave propagates in the sample along the interface, and we have seen that the wave intensity decreases exponentially with the distance from the interface. If the sample absorbs, the absorption of the evanescent wave occurs proportionally with the intensity of the evanescent wave. Most of the electromagnetic energy is absorbed by the layer of the sample closest to the interface. If we imagine that the sample consists of layers parallel to the interface, then the amount of light absorbed in each layer is governed by the absorption index of the sample and the intensity of the evanescent wave in the layer. From Equation 7.9, for ATR reflectance and by substituting the expression (Eq. 5.4) for penetration depth, we get

$$\frac{P_A}{P_{in}} = 1 - R^{s,p} = \frac{n\kappa\left|t_{12}^{s,p}\right|^2}{n_1 \cos\theta\sqrt{n_1^2 \sin^2\theta - n^2}} = \frac{4\pi n\kappa k}{n_1 \cos\theta}\left|t_{12}^{s,p}\right|^2 \int_0^\infty e^{-\frac{2z}{d_p}} dz. \quad (10.1)$$

The above expression was derived under the assumption that the sample is uniform so that the optical constants of the sample could be taken outside the integral. However, if the sample is not uniform and the optical constants change with the depth into the sample, they have to be left under the integral. Since the real part of the refractive index for most samples is approximately the same, the changes in the absorption index will dominate the above expression. Thus, in the first approximation, we need only to retain the absorption index under the integral. The resulting term under the integral is thus

Internal Reflection and ATR Spectroscopy, First Edition. Milan Milosevic.

$$I(d_p) = \int_0^{\infty} \kappa(z) e^{-\frac{2z}{d_p}} dz. \tag{10.2}$$

The expression (Eq. 10.2) is essentially the Laplace transform of the absorption index. This opens up an intriguing possibility that, at least in principle, one can acquire a number of spectra at different angles of incidence, hence different penetration depths, and experimentally measure $I(d_p)$. The hope is that one can then use the experimentally measured $I(d_p)$ to calculate the inverse Laplace transform and therefore recover how the absorption index changes with depth.

The main difficulty with such a program is that the experimentally accessible range of values for d_p is not wide. The d_p for one can never be zero. Even at the angle of incidence of 90°, there is a residual penetration depth. On the other side of the angular range, for the angle of incidence near the critical angle, the depth of penetration is an extremely sensitive function of the angle of incidence and is diverging to infinity at the critical angle. In the spectral regions where the sample absorbs, the concept of penetration depth loses its clear meaning and needs to be modified. Thus, a relatively narrow range of penetration depths can be used to measure $I(d_p)$. Another complicating factor with depth profiling in ATR is that penetration depth is a function of wave number (wavelength).

However, all this does not imply that no information on the variation of the absorption index with depth can be obtained in this way. Since the integral in Equation 10.2 is essentially a measurable quantity, we can evaluate Equation 10.2 in some important practical cases. First, we assume that $n = 1.5$ and that the absorption index κ is constant throughout the sample. Then,

$$I(d_p) = \kappa \frac{d_p}{2}. \tag{10.3}$$

The expression (Eq. 10.3) states the obvious; that is, everything else being the same, the absorbance will increase in proportion to the penetration depth.

10.2 THIN ABSORBING LAYER ON A NONABSORBING SUBSTRATE

Another important and common case is the one where the absorption index is constant within a thin layer of thickness d and vanishes in the rest of the sample. A typical case would be that of a thin coating over a substrate. In that case, Equation 10.2 evaluates to

$$I(d_p) = \kappa \frac{d_p}{2}\left(1 - e^{-\frac{2d}{d_p}}\right). \tag{10.4}$$

The expression (Eq. 10.4) starts out just as Equation 10.3, which should not be surprising. For a penetration depth much shorter than film thickness, the situation is similar to that for an infinitely thick film. ATR cannot distinguish between two different thicknesses of the layer so long as they are both much larger than the penetration depth. At the penetration depths much larger than the film thickness, the integral saturates and no longer increases.

10.3 THIN NONABSORBING FILM ON AN ABSORBING SUBSTRATE

The third case of general interest is when the film is not absorbing but the substrate absorbs. This is the reverse of the previous case. In this case, the integral (Eq. 10.2) evaluates to

$$I(d_p) = \kappa \frac{d_p}{2} e^{-\frac{2d}{d_p}}. \tag{10.5}$$

The expression (Eq. 10.5) starts out by not increasing at all with an increasing penetration depth. Then, after the penetration depth approaches the film thickness d, the integral takes off and eventually settles into a linear growth with the same slope as in the first case. Again, this is intuitively clear since, for very short penetration depths, the evanescent wave is entirely confined to the nonabsorbing film. It is only after the penetration depth approaches the film thickness that there is a portion of the evanescent wave that extends into the absorbing substrate and is there absorbed. The graphs of the three cases are shown in Figure 10.1.

10.4 THIN NONABSORBING FILM ON A THIN ABSORBING FILM ON A NONABSORBING SUBSTRATE

Another simple, but practically important, case is one of the nonabsorbing film of thickness d_1 on the absorbing film of thickness d_2 over a nonabsorbing substrate. Figure 10.2 shows the case in which $d_1 = 2\ \mu\text{m}$

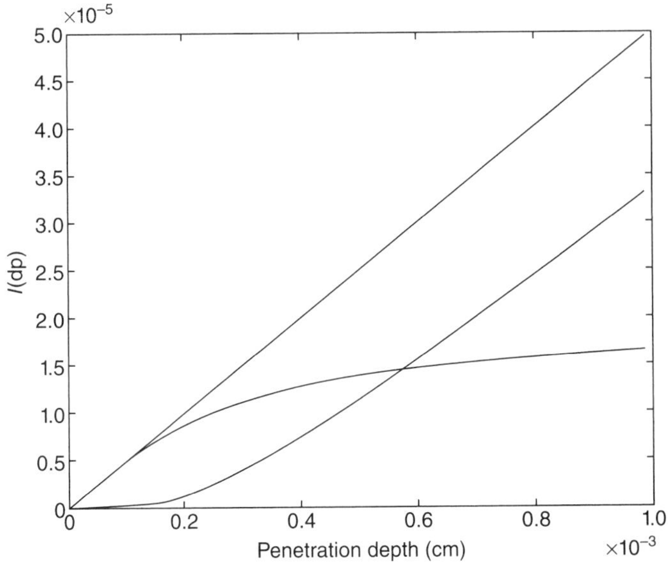

Figure 10.1 $I(d_p)$ versus d_p for the three cases discussed in the text.

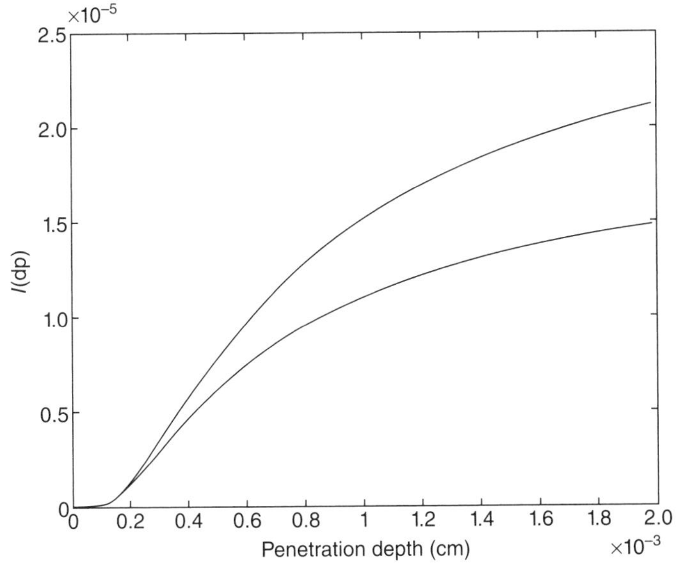

Figure 10.2 Depth profiles for 5 μm (lower) and 6-μm-thick absorbing layers on a nonabsorbing substrate and covered by a 2-μm-thick nonabsorbing layer.

and $d_2 = 5\ \mu m$. Figure 10.2 also shows a similar case with $d_1 = 2\ \mu m$ and $d_2 = 6\ \mu m$ to illustrate what can possibly be resolved using the depth profiling technique. As we can see, both graphs show no absorption at very shallow penetrations, and both saturate at deeper penetrations. In all cases displayed in Figures 10.1 and 10.2, the value of the absorption index used was $\kappa = 0.1$.

We chose the same value of the real part of the refractive index for the absorbing and nonabsorbing materials in order to avoid having to treat the interface in a rigorous way. However, the absorption index had a clear discontinuity at the interface. It is therefore not at all clear that the interface could be ignored. There could be a Fresnel reflection even if only the absorption index changes significantly at the interface.

On the other hand, we here have an evanescent wave that propagates parallel to the interface, so how could such a wave reflect from the interface? Thus, if there can be no reflection at the interface, it would seem justified to ignore any effect it may cause. However, Fresnel equations link the fields of the electromagnetic waves on the two sides of the interface, so the interface does have an effect that was not taken into account in the above analysis.

The interpretation of the depth profile curves becomes more complicated when the absorption index of the sample is not known. Note that it would be hard to distinguish if the higher curve in Figure 10.2 is higher because the absorbing layer is thicker than the one for the lower curve, or because the two layers are equally thick, but that the absorption index is stronger for the higher curve.

The cases shown suggest that, in general, it should be possible to deduce at least the rough features of the depth profile for an unknown sample, but they also suggest the difficulties that such an analysis confronts. In general, if one has a good idea what a depth profile for a particular sample may be, the depth profiling analysis may be very helpful in extracting the parameters such as the layer thicknesses or the refractive indices of the different layers.

11 Multiple Interfaces

11.1 REFLECTANCE AND TRANSMITTANCE OF A TWO-INTERFACE SYSTEM

In the previous chapter, we studied attenuated total reflection (ATR) reflectance of samples that are not homogeneous in their composition. A typical example would be a sample that has its surface chemically modified or a sample with a thin film coating applied over the surface. A number of samples of this type are of considerable practical importance. We have, however, assumed that the interfaces separating different layers in the sample are not contributing in a significant way to the measured reflectance. Now, we are turning our focus to the case where multiple interfaces are present. We first consider the case of three media with refractive indices n_1, n_2, and n_3 separated by two parallel interfaces a distance d apart.

Let light be incident from the first medium at the interface between media 1 and 2. As usual, we assume that the incident medium does not absorb. Media 2 and 3 could be, but do not have to be, absorbing. Incident light will both transmit the first interface and will reflect from it. Transmitted light propagates through medium 2 until it hits the second interface. There it both reflects and transmits. The transmitted light continues into medium 3. The reflected light travels backward through medium 2 back toward the first interface. When it reaches the first interface, it again both reflects and transmits, and so on. There is an infinite sequence of reflections between the two interfaces. The reflected component keeps reflecting back and forth, while the transmitted component contributes either to transmitted light or to reflected light depending whether the reflection occurs on the first or on the second interface. The situation is sketched in Figure 11.1.

The situation shown in Figure 11.1 is pretty much what would have been seen if laser light was shone onto the first interface. However, we envision the incident plane wave to have an infinitely extended wave

Internal Reflection and ATR Spectroscopy, First Edition. Milan Milosevic.

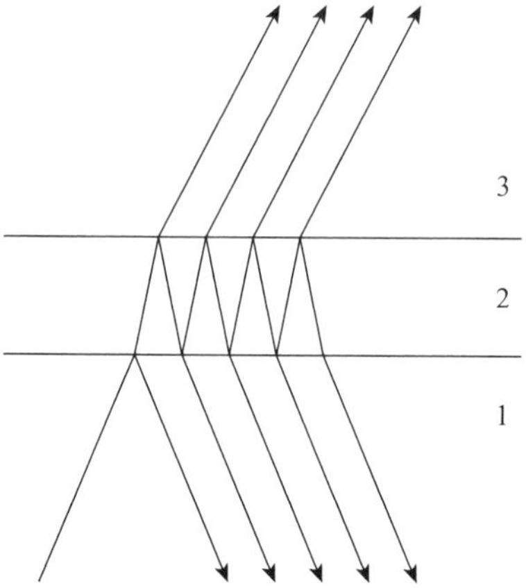

Figure 11.1 Multiple reflections at a two-interface system.

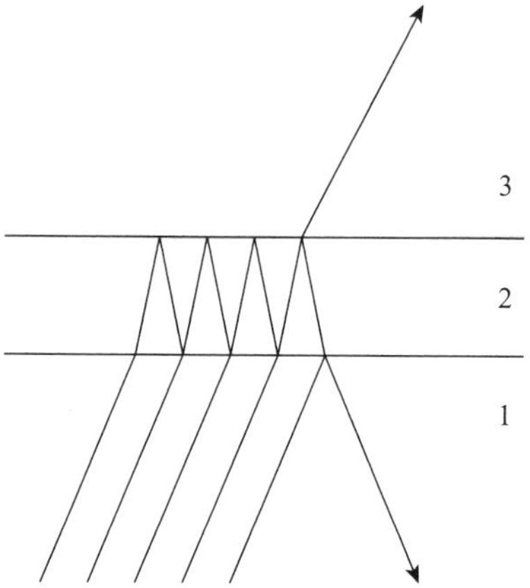

Figure 11.2 Alternative view of multiple reflections at a two-interface system.

front and every point on this wave front to be moving along a trajectory such as shown in Figure 11.1.

The multiple waves contributing to the reflected light are displaced along the interface so they cannot contribute to the wave amplitude in the same point on the interface. Instead of the situation shown in Figure 11.1, we should consider a more appropriate depiction shown in Figure 11.2. Here we choose the points on the wave front that arrive at the first interface earlier so that all the reflected waves coincide at the same point contributing to the reflected wave amplitude at that point.

The amplitude of the reflected wave is a sum of infinitely many waves. It is relatively straightforward to sum all the waves. For each multiply reflected wave, we count the number of reflection and transmission amplitude coefficients picked up by the wave zigzagging between the two interfaces and add the phase factor that describes multiple passes between the two interfaces. Since all the multiple components add to the amplitude of the reflected wave in the same point on the first interface, the only difference in the phase factor between all the components is contained in the z-axis factor, which accounts for how many times a particular component went from the first interface to the second interface and back. As always, we assume that the z-axis is perpendicular to the interfaces. The z-axis factor in the exponential term that describes propagation of the plane wave is

$$e^{2\pi i n_2 k z \cos\varphi}. \tag{11.1}$$

On its way to the second interface ($z = d$), the wave picks up a propagation factor (Eq. 11.1), and then on its way back to the first interface ($z = 0$), it picks up another one. The result is

$$\begin{aligned} A_r &= \left(r_{12} + t_{12}r_{23}t_{21}e^{4\pi i n_2 k d\cos\varphi} + t_{12}r_{23}r_{21}r_{23}t_{21}e^{4\pi i n_2 k d\cos\varphi} + \cdots\right)A_{in} \\ &= r_{12}A_{in} + t_{12}r_{23}t_{21}e^{4\pi i n_2 k d\cos\varphi}\left(1 + r_{21}r_{23}e^{4\pi i n_2 k d\cos\varphi} + \cdots\right)A_{in} \\ &= \left(r_{12} + \frac{t_{12}r_{23}t_{21}e^{4\pi i n_2 k d\cos\varphi}}{1 - r_{21}r_{23}e^{4\pi i n_2 k d\cos\varphi}}\right)A_{in} = \frac{r_{12} + r_{23}e^{4\pi i n_2 k d\cos\varphi}}{1 + r_{12}r_{23}e^{4\pi i n_2 k d\cos\varphi}}A_{in}, \end{aligned} \tag{11.2}$$

where we used $r_{21} = -r_{12}$ and $t_{12}t_{21} + r_{12}^2 = 1$, and we summed up the infinite series by using the formula for geometric series. The result (Eq. 11.2) is applicable to both s and p polarizations.

We can use the same procedure to obtain the amplitude of the transmitted wave and we find

$$\begin{aligned} A_t &= \left(t_{12}t_{23}e^{2\pi i n_2 k d\cos\varphi} + t_{12}r_{23}r_{21}t_{23}e^{6\pi i n_2 k d\cos\varphi} + \cdots\right)A_{in} \\ &= \frac{t_{12}t_{23}e^{2\pi i n_2 k d\cos\varphi}}{1 + r_{12}r_{23}e^{4\pi i n_2 k d\cos\varphi}}A_{in} \end{aligned} \tag{11.3}$$

In the above expressions, we can use Snell's law to replace

$$n_2\cos\varphi = \sqrt{n_2^2 - n_1^2\sin^2\theta}. \tag{11.4}$$

In analogy with the definition of single interface reflection and transmission amplitude coefficients, we can define the reflection and transmission amplitude coefficients for the two-interface system as

$$\tau = \frac{A_t}{A_{in}}, \quad \rho = \frac{A_r}{A_{in}}, \tag{11.5}$$

with the definition valid for both polarizations. Obviously, the same definition can be extended to a system consisting of an arbitrary number of interfaces. With the definitions in Equations 11.4 and 11.5, the system consisting of two interfaces can be formally replaced with a generalized interface having the reflection and transmission amplitude coefficients:

$$\rho = \frac{r_{12} + r_{23} e^{4\pi i k d \sqrt{n_2^2 - n_1^2 \sin^2\theta}}}{1 + r_{12} r_{23} e^{4\pi i k d \sqrt{n_2^2 - n_1^2 \sin^2\theta}}} \tag{11.6a}$$

$$\tau = \frac{t_{12} t_{23} e^{2\pi i k d \sqrt{n_2^2 - n_1^2 \sin^2\theta}}}{1 + r_{12} r_{23} e^{4\pi i k d \sqrt{n_2^2 - n_1^2 \sin^2\theta}}}. \tag{11.6b}$$

In a way, these generalized coefficients are very much similar to the single interface amplitude coefficients and allow us to treat a two-interface system as a single interface. However, the structure of these generalized coefficients is much richer than the structure of the single interface coefficients. The Fresnel amplitude coefficients r and t have a rich structure as we have already seen. Now, in addition to these coefficients, the exponential factor containing the distance between the two interfaces (i.e., the thickness of the layer) and the wavelength of light is explicitly present in the expressions (Eq. 11.6). The exponential term contained in these expressions exhibits oscillatory behavior for the subcritical angles of incidence.

In reflection experiments, the reflected light returns back to the incident medium. So, medium 3 can be, and often is, different from medium 1. In transmission experiments, on the other hand, the incoming medium 1 and the exiting medium 3 are almost always the same.

11.2 VERY THIN FILMS

Note also that if the thickness d of medium 2 becomes zero, the parameters of medium 2 should disappear from the above expressions. The system of two interfaces transforms into a single interface between media 1 and 3. Consequently, as $d \to 0$, we have $\rho \to r_{13}$ and $\tau \to t_{13}$. Thus, if media 1 and 3 are the same, the reflection coefficient ρ becomes zero and the transmission coefficient τ becomes one. It is intuitively obvious that it should be so since by letting $d \to 0$, the two interfaces merge and medium 2 disappears.

On first sight, this result appears puzzling. After all, the reflectance coefficient of each interface is independent of other interfaces or distances between them. For normal incidence, the reflectance coefficient is only a function of the refractive indices of the two media forming the interface (Eq. 4.17a). So how is it possible that the reflectance of the two-interface system disappears as the distance between the interfaces tends to zero? The answer is that the reflectance amplitude coefficients at the first interface r_{12} and at the second interface r_{21} are of the opposite sign so the reflected wave from the second interface interferes destructively with the wave reflected from the first interface and the intensity of the reflected wave is thus zero. This is strictly true only if the thickness of the film is negligible with respect to the wavelength so the wave reflected at the second interface is not appreciably delayed in phase behind the wave reflected at the first surface. Of course, as the reflectance of such a thin film vanishes, the transmittance becomes total and light propagates through the film as if it were not there.

11.3 INTERFERENCE FRINGES

As the thickness of the film increases, the beam reflected from the second interface starts lagging in phase behind the beam reflected at the first interface. The cancellation that occurred when the thickness was very close to zero is no longer perfect. As the optical path difference increases further and reaches half a wavelength of the reflected light, the beam reflected at the first interface and at the second interface come perfectly in phase, reinforcing each other and bringing the intensity of the reflected beam to maximum. As the thickness increases further, the reflectance starts decreasing reaching again minimum in reflected intensity as the optical path difference equals the wavelength of light, and so on. The transmittance of light through the film similarly oscillates with thickness. The transmittance is at maximum for the film thickness at which the reflectance vanishes and it is at minimum when reflectance is at maximum. These are the well-known interference fringes. A thin film of oil on the surface of water is a familiar example from everyday life that exhibits interference fringes in reflection.

In spectroscopy, the fringes manifest themselves in a slightly different but related way. Instead of observing the fringes as the film thickness changes, the fringes are measured for a thin film of a given thickness as a function of wavelength. Instead of film thickness, the wavelength of light changes, and minima and maxima in reflection or transmission occur as oscillating functions of the wavelength (wave number).

If the film does not absorb light, then light is either reflected or is transmitted. The conservation of energy requires that the transmittance and reflectance of a nonabsorbing film add to one.

The phenomenon of interference fringes has an enormous range of applications. One important application is to coat a thin film of appropriate material over a surface of an optical element in order to reduce or even eliminate reflection loss that would have otherwise occurred at the surface of the element. We have seen how this would work for a single wavelength. We would have to choose the film thickness that for that wavelength causes the component reflected from the second interface of the film to interfere destructively with the component reflected from the first surface of the film.

Appropriate selection of the refractive index of the film such that the reflectance amplitude coefficients at the two interfaces are of the same magnitude eliminates reflected light and thus yields lossless transmission through the interface. A careful design of a multilayer coating can broaden this antireflection coating to work over a range of wavelengths. Antireflection coatings covering the entire visible spectral range are commonplace on quality lenses. Antireflection coatings covering the infrared (IR) spectral region from about 2.5 μm to about 16 μm are routinely applied to high refractive index IR materials such as ZnSe, Ge, and Si.

The film does not have to be thin for interference fringes to form. Two plane parallel partially reflective interfaces generate interference fringes at any separation. The frequency of oscillations as a function of wave number increases with the distance. Thus, for very thick films, the fringes are very close together and are difficult to resolve.

This phenomenon of interference fringes impacts the measurements of the transmittance of thin samples. In IR, the absorbance of most samples is very strong and the path length required is typically in tens to low hundreds of micrometers. Such path lengths produce interference fringes that strongly obscure the measured reflectance of the sample. There are a number of tricks that are used to reduce the amplitude of fringes, but they all come at the expense of blurring some other experimental parameter such as the path length.

11.4 NORMAL INCIDENCE

To illustrate the phenomenon of interference fringes, let us consider a case of a nonabsorbing thin film. A good and very common example would be an empty liquid cell. To analyze liquids in transmission, a cell is typically made by injecting liquid between two optically transparent disks (windows) held at a fixed and known separation by a spacer. To remove the effect of window absorption, the cell is sometimes placed into the spectrometer empty; that is, the space between the windows normally occupied by the liquid sample is left empty. Thus, two reflecting surfaces (inner faces of the windows) are placed close together, and that, as we have seen, leads to interference fringes. If the incidence is normal as it almost always is ($\theta = 0$), the sample is air ($n_2 = 1$) and materials 1 and 3 are the same, the expression (Eq. 5.10) reduces to

$$\tau = \frac{\left(1 - r_{12}^2\right) e^{2\pi ikd}}{1 - r_{12}^2 e^{4\pi ikd}}. \tag{11.7}$$

The transmittance is given by the absolute value squared of the above expression. Thus,

$$T = |\tau|^2 = \frac{1}{1 + F\sin^2 2\pi kd}, \tag{11.8}$$

where

$$F = \frac{4|r_{12}|^2}{\left(1 - |r_{12}|^2\right)^2} \tag{11.9}$$

is the so-called *finesse*. For normal incidence,

$$|r_{12}|^2 = \left(\frac{n-1}{n+1}\right)^2. \tag{11.10}$$

For a typical optical material (such as glass), the refractive index is around 1.5, so the reflectance of the interface is around 0.04 (4%). Thus, a typical finesse is around 0.16.

Starting from expression in Equation 5.10 and following similar steps as above, it is easy to see that the normal reflectance of the film is

$$R = \frac{F\sin^2 2\pi kd}{1 + F\sin^2 2\pi kd}. \tag{11.11}$$

The transmittance (Eq. 11.8) and reflectance (Eq. 11.11) are undulating functions of wave number with the period:

$$\Delta k = \frac{1}{2d}. \tag{11.12}$$

The maxima of transmittance occur for those wave numbers for which the sine is zero. The minima occur where sine squared is one. Thus, T oscillates between the values of 1 and $1/(1 + F)$. The amplitude of those oscillations is relatively small for a typical refractive index. However, if the reflectivity of the interface is enhanced by evaporating a thin film of metal, high finesse values can be achieved. With a high finesse, the oscillations in transmission become very strong and the peaks at $T = 1$ become very narrow. It is said that the fringes have high contrast.

Note that $T + R = 1$ as a quick inspection of the expressions in Equations 11.8 and 11.11 would confirm. Therefore, the fringes have nothing to do with light absorption. Minimum of reflectance simply indicates that light of that particular wavelength has been maximally transmitted.

However, in the case of an absorbing sample, $T + R < 1$. All the light that is neither reflected nor transmitted is absorbed. The fraction of light that is absorbed in the film is thus equal to $1 - T - R$.

A very useful practical application of observing interference fringes under these conditions is that it provides a very precise measurement of the cell's path length (i.e., the separation between the windows). By taking a transmission spectrum of such a cell and finding the period of oscillations, the cell's path length is easily found from Equation 11.12.

Another useful application of this phenomenon is that a very precise measurement of distance can be achieved by sending a laser beam through this setup and by counting fringes on the laser detector as the distance between the windows changes. In this application, a higher finesse is helpful to increase the amplitude of fringes. Each time the distance between the windows changes by half of the wavelength of the laser light, the fringe counter adds another count. Thus, a length could be measured to within a fraction of wavelength of the laser light.

11.5 INTERFERENCE FRINGES AND TRANSMISSION SPECTROSCOPY

As useful as the phenomenon of interference fringes is in other applications, it is a nuisance in measuring the transmittance of a sample. Figure 11.3 illustrates the type of difficulties arising from the phenomenon of interference fringes. The spectra shown are simulated using the expres-

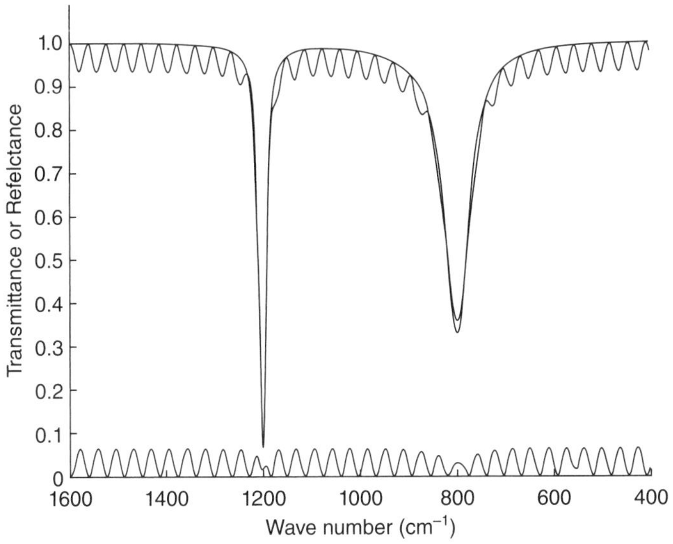

Figure 11.3 Absorption peaks in the reflectance and transmittance spectra and the sum of the two spectra.

sions in Equation 11.6 and the harmonic oscillator model described earlier. The thickness of the sample was chosen as 0.100 mm, and the real part of the refractive index, away from the absorbing regions, was chosen to be a typical 1.5.

The reflectance spectrum barely reveals the absorption peaks. The peaks show as minor disturbances to the otherwise uniform oscillating reflectance curve. The transmittance spectrum shows absorption peaks more clearly, but it is still hard to decide where the baseline is, and therefore it is hard to precisely determine the height of the absorption peak—a necessary ingredient for quantitative spectroscopy. In addition, the peak shapes are affected so it is not easy to delineate if the shoulder on the right side of the peak at 1200 cm^{-1} is a true absorption peak or if it is an artifact due to interference fringes. Weaker peaks may be completely obscured.

The sum of the two spectra (also shown) represents the fraction of incident light that was not absorbed by the sample. Therefore, the absorption peaks seen in the sum spectrum directly quantify the fraction of light absorbed in the experiment.

Interference fringes are largely removed by the addition. However, a remnant of the fringes is still visible around the shoulders of the peaks. The reason for these remaining undulations is that the fringes modulate how much light intensity propagates back and forth between the two interfaces. The more light intensity is available, the more is absorbed.

The maxima of light intensity between the two interfaces line up with the maxima seen in the reflectance spectrum. Therefore, the interference fringes are not just a superficial artifact; they actually influence the observed absorbance. We also see that the peak at 800 cm^{-1} is significantly affected by the fringes. Therefore, not just the baseline but the peak itself is obscured by the presence of fringes.

The only way to remove the effect of fringes on the transmittance of the film would be to somehow make the reflectance of the film surfaces vanish. Although making the reflectance of film surface vanish may seem impossible, there are actually two ways in which that can almost be accomplished.

The first way is to interface the film on both sides with an optically transparent medium of a refractive index that matches that of the film. Figure 11.4 shows the sample of Figure 11.3 but now immersed into a material with the refractive index equal to 1.50. Since light is incident onto the film from the material with the refractive index that is matched to that of the sample, the interfaces cease being reflective, and thus there are no multiple reflections within the film. Note that either the incident or the final medium needs to be refractive index matched to the sample in order to remove multiple reflections inside the film.

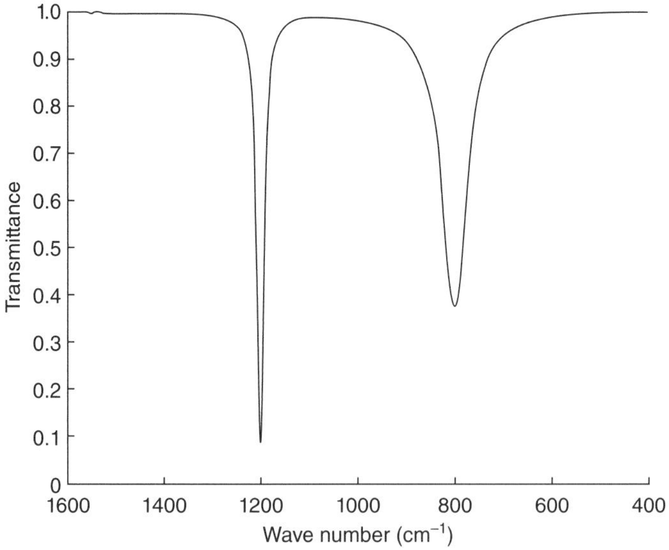

Figure 11.4 The transmittance spectrum of a sample immersed into a medium of a matching refractive index.

However, if the matching of refractive indices is not perfect, it is advantageous to use the same index matching material for both the first and the last medium.

We can see from Figure 11.4 that the fringes are removed. However, a close inspection reveals that a remnant of fringes is still present. This is obviously due to the imperfect match between the refractive indices of the two media. A perfect match is impossible since the refractive index of the sample is significantly affected near the absorption bands.

Another way of nearly eliminating interference fringes is to use *p*-polarized light and to choose the angle of incidence to be Brewster's angle. We have seen that, at the Brewster's angle, the reflectance of *p*-polarized incident light vanishes. Again, just as with the case of immersing the sample into a medium of matching refractive index, the changes in the refractive index of the sample, in particular near the absorption bands, make it impossible to eliminate reflectance for the entire spectral range. However, by choosing an angle that minimizes the fringes, a significant improvement to the transmission spectrum can be achieved. The transmittance of the sample from Figures 11.3 and 11.4 is shown in Figure 11.5.

Notice that the absorption peaks shown in Figure 11.5 are stronger than those shown in Figure 11.4. The reason for this is that the path length through the film in Figure 11.5 is now longer than 100 μm because light travels through the film at an angle.

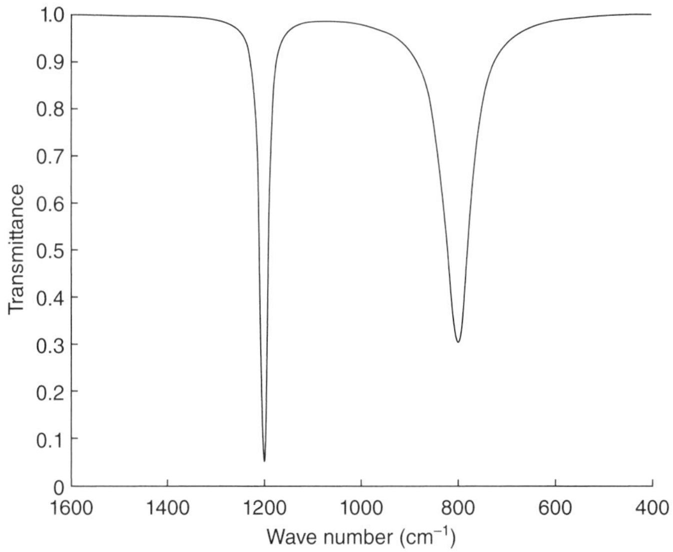

Figure 11.5 The transmittance of a sample for *p*-polarized light incident at 55.6° (Brewster angle).

11.6 THIN FILMS AND ATR

As we have mentioned earlier, the phenomenon of interference fringes is absent in ATR even if the sample is a thin film. The ATR spectrum of the same sample used above is shown in Figure 11.6 for p-polarized light, at 45° angle of incidence, and for the ATR material with a refractive index of 2.42. The peak strengths are much weaker than in transmission since the effective path length for the ATR spectrum is much shorter than the path length in transmission.

Two spectra are shown, one for a sample thickness of 100 μm and the other for a sample thickness of 3 μm. For the thicker sample, the penetration of the evanescent wave into the sample material is much shorter than the thickness of the sample and the presence of the second interface is irrelevant to the evanescent wave. However, even if the film thickness is much smaller and the second interface is within the reach of the evanescent wave, as it is for the 3-μm-thick film, the interference fringes are still completely absent from the ATR spectra.

The spectrum of the 3-μm-thick film shows absorption peaks that are just slightly weaker than they are for the 100-mm-thick sample. Thus, clearly, the evanescent wave reaches the top surface of the thinner sample and extends into the nonabsorbing air behind the film. Since a portion of the evanescent wave travels through the nonabsorbing air,

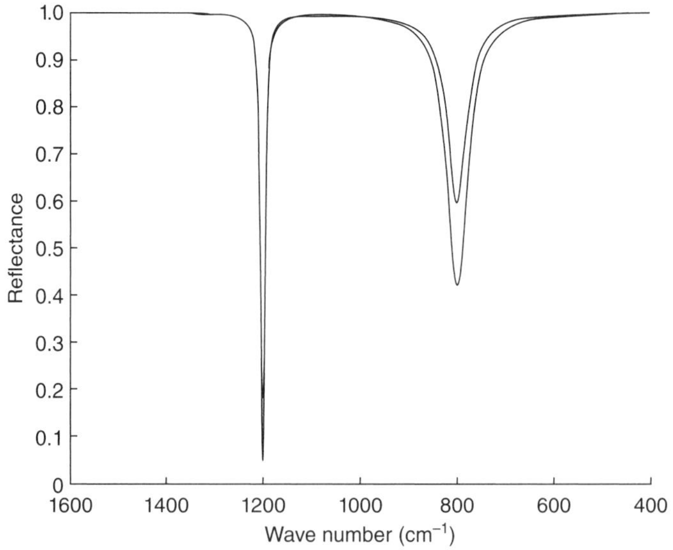

Figure 11.6 ATR spectra of a 100- and 3-μm-thick sample (p polarization at 45° angle of incidence).

the absorption peaks are less intense for the 3-μm-thick sample. Note, however, that the fact that the absorption peaks for the 3-μm-thick sample are not that much weaker than for the 100-μm sample implies that the evanescent wave does not penetrate much deeper than 3 μm into the sample. Despite the fact that two closely spaced interfaces are present, the interference fringes are absent from the spectrum.

11.7 INTERNAL REFLECTION: SUBCRITICAL, SUPERCRITICAL, AND IN BETWEEN

The case of internal reflection from a thin film on an ATR crystal can be split into the three ranges with regard to the angle of incidence (Fig. 11.7). The first range corresponds to an angle of incidence between the normal and critical angle θ_{c1} for the ATR material and air ($n_3 = 1$):

$$\sin\theta_{c1} = \frac{n_3}{n_1}. \tag{11.13}$$

The second range is between the critical angle θ_{c1} and the critical angle between the ATR material and the sample θ_{c2}:

$$\sin\theta_{c2} = \frac{n_2}{n_1}. \tag{11.14}$$

The third range is for angles above the critical, that is, for $\theta > \theta_{c2}$.

The spectra in the first range of angles of incidence are related to the reflection spectra of a freestanding thin film. The spectrum of 100-μm-thick film deposited on an ATR material with refractive index $n = 2.42$ is shown in Figure 11.8. The material is the same as the one shown in Figure 11.3. Note that the interference fringe pattern is very similar to that of the freestanding film. The fringe frequency is the same. There are two disturbances in the regular undulating pattern where the absorption peaks are. The fact that the reflectance of the

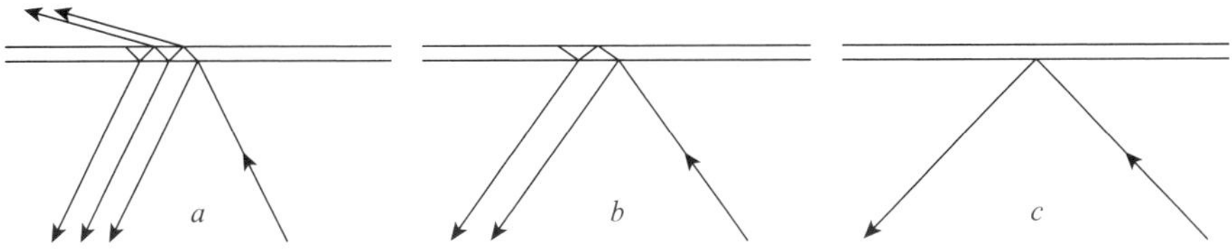

Figure 11.7 Three types of internal reflection from a thin film.

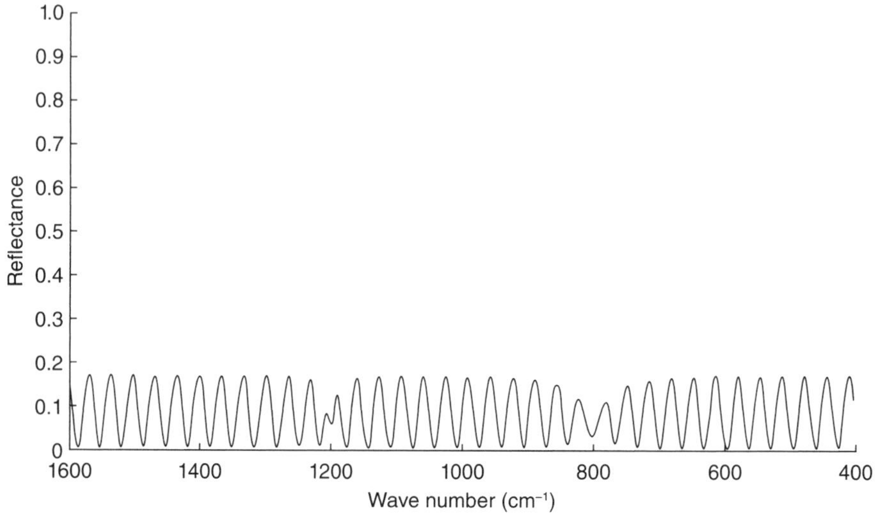

Figure 11.8 The normal incidence reflectance spectrum of a 100-mm-thick film deposited on a material of a refractive index of 2.42.

two interfaces of the film are not equal as they are for the freestanding film does not seem to change anything substantial. Note that the fringe amplitude is stronger in this case, reflecting the fact that the reflectance of the first interface is now larger than the reflectance of the second interface.

11.8 UNUSUAL FRINGES

An interesting and not as yet generally recognized as a distinct phenomenon is the reflection that occurs in the range of angles of incidence between θ_{c1} and θ_{c2}. In this angular range, the first interface is not totally reflecting. In addition to the component reflected from the first interface, there is also a component of light that transmits through the interface and propagates into the film. At the second interface, this component is totally internally reflected so none of the light escapes the film. Light totally internally reflected from the second surface returns back to the first interface where a part of it transmits through the interface and a part reflects back into the film, and so on.

Note that light propagates through the film as a propagating wave, not as an evanescent wave. Thus, the path length through the film could be very long. Figure 11.9 illustrates the complexity that this mode of internal reflection can produce. The sample is the same 100-μm-thick

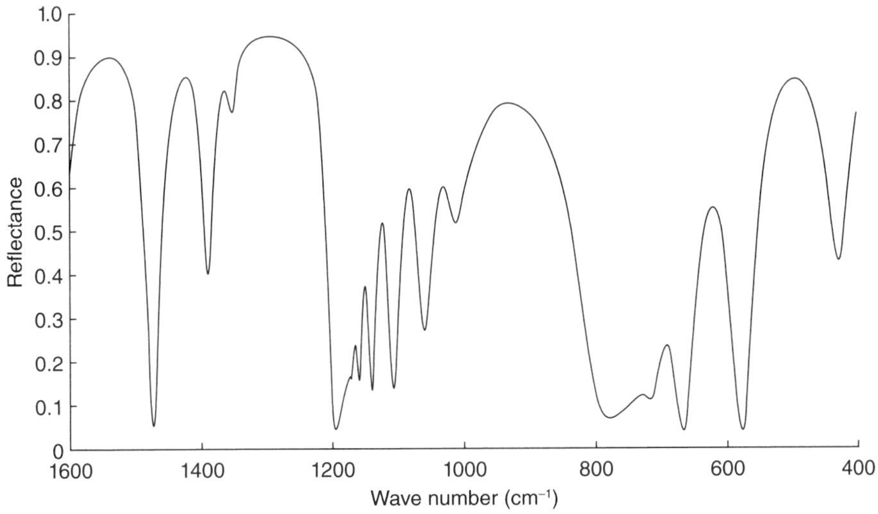

Figure 11.9 Internal reflectance of a 100-μm-thick film.

film shown in Figure 11.6. Figure 11.9 shows the reflectance of *p*-polarized light at the angle of incidence of 37.2°. A striking feature of this spectrum is the incredible complexity apparently unrelated to the simple ATR spectrum of the same sample shown in Figure 11.6. A large number of peaks are now present. Unlike the interference fringes, which are all uniform and equally spaced, these peaks appear randomly positioned and are of different strengths and widths. Note a number of strong peaks that are not at all the absorption peaks of the sample material. The actual absorption peaks of the material positioned at 1200 and 800 cm^{-1} are clearly present, but their lower wave number side is dotted with a number of satellite peaks.

The appearance of these satellite peaks is puzzling. As we have already mentioned, with interference fringes, the energy of the incident beam is split into transmitted and reflected beams. Depending on the interference mechanism which is controlled by the wavelength of light, the angle of incidence, and the thickness of the film, either the transmission or the reflection channel is favored. We see fringing in both channels, but as we have seen above, if the two channels are added together, the fringing is considerably suppressed. As we have also seen, however, a weak remnant of fringing is still present in the sum spectrum. This remaining fringing in that case is a consequence of the enhanced radiation intensity inside the film that is, at some wavelengths, brought about by the constructive interference of the multiple reflected beams inside the film. For those wavelengths for which the radiation intensity inside

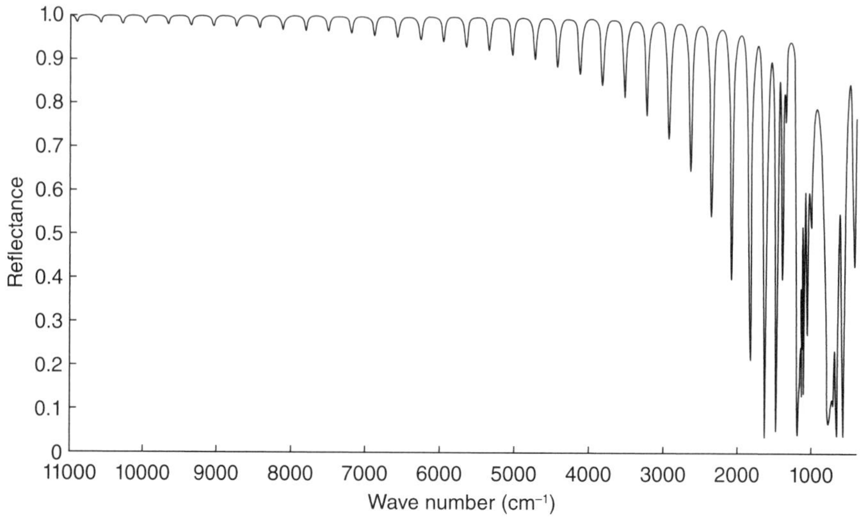

Figure 11.10 Internal reflectance of the 100-μm-thick film from Figure 11.9 shown over a much broader spectral range.

the film is increased, the absorption of light by the medium is proportionally increased and conversely for the destructive interference. Hence, tiny remnant fringes remain even when the reflection and transmission channels are added together.

However, in the present case, the transmission channel is not present, so the reflected beam contains all the energy carried by the incident radiation that was not absorbed in the sample. Figure 11.10 illustrates the extent of this phenomenon by showing the same spectrum from Figure 11.9, but in a much wider spectral range. Note the large number of additional peaks that were not visible in Figure 11.9. These peaks are almost equally spaced, but they are more closely spaced and of increasing intensity near the absorption peaks. Further away from the absorption bands, they are more regularly spaced and somewhat resemble interference fringes.

The question is what is causing these fringes? Where did the light missing in the reflected beam go? Normally, with interference fringes, the light missing in the reflected beam either appears in the transmitted beam or was absorbed in the sample. We have seen that, with the absorption peaks present in the normal incidence spectrum of a thin film, the addition of the reflected and transmitted beams drastically reduces the interference fringes in the sum spectrum, although it does not fully eliminate them.

However, if the sample is nonabsorbing, no fringes are present in the sum spectrum. In the present case, there is no transmitted beam. All the light, not absorbed by the sample, must be contained in the reflected beam. The fringes stretch far from the absorption peaks. The only way the peaks could occur at all is if the light is absorbed by the sample. So the peaks that we see are both the interference fringes and the absorption peaks, but a type of absorption peak that is not the absorption peaks characteristic of the sample. The absorption index of the sample is virtually negligible about 1000 cm^{-1} or more away from the absorption peaks. So, the fact that there are strong absorption fringes/peaks where the absorption index of the sample is so incredibly small implies that there is an enormously powerful resonance mechanism that boosts the absorption of light at these wavelengths.

The positions of the peaks are determined entirely by the thickness of the film, its refractive index, and the angle of incidence and are not at all a consequence of the residual absorption of the material. The fringes are very sensitive to the sample thickness and to the angle of incidence. A miniscule variation in either of them considerably impacts the position of the fringes. Thus, to experimentally observe this phenomenon, an incident beam with a very low divergence, such as a laser beam, would have to be used. This phenomenon clearly demonstrates that there are still unexplored phenomena hidden inside the well-known and deceptively simple expressions (Eq. 11.6).

11.9 PENETRATION DEPTH REVISITED

We have already studied the penetration of the evanescent wave into a sample in contact with a totally internally reflecting interface. We have also recognized the potential to selectively probe the sample at different depths from the interface by varying the penetration depth. However, we have assumed that the changing composition of the sample does not impact the evanescent wave in a significant way and that we can use the form of the evanescent wave that was derived under the assumption that the medium is homogeneous. In particular, when we considered a two-layer system, we ignored the presence of the interface between the two layers. It is not clear which role such an interface could even play for an evanescent wave since the wave propagates parallel to the interface. We expect, however, that, since Fresnel amplitude coefficients connect fields on the two sides of the interface between the layers, the interface must affect the evanescent wave in some way. We can now revisit the penetration depth analysis this time using the description based on the two-interface system.

One way to analyze the penetration of the evanescent wave beyond the totally internally reflecting interface is to place an air gap between the interface and the sample. If the thickness of the air gap is varied and a change in the strength of an absorption band of the sample is observed, a clear picture of the extent of penetration of evanescent wave into the air gap can be seen. To illustrate, we chose a ZnSe-like ATR prism, a 45° angle of incidence, and *p*-polarized incident light. We used our standard sample and turned off all the other peaks except the one at 1000 cm^{-1}.

The penetration depth, of course, is influenced by the refractive index of the material through which the evanescent wave propagates, and by replacing the air gap with a nonabsorbing medium of the same refractive index as the sample and the same thickness as the air gap, the depth of penetration increases considerably. The two cases are shown in Figure 11.11.

Note that the two curves start at the same point for zero thickness of the nonabsorbing layer. As we would expect, they both decay expo-

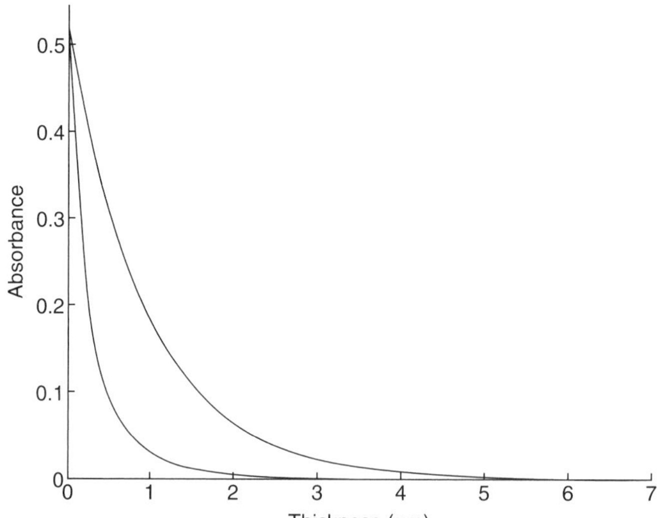

Figure 11.11 Absorbance of the sample spaced from the ATR interface by a nonabsorbing layer as a function of the layer thickness. The lower curve corresponds to the case where the nonabsorbing layer is air. The higher curve shows this dependence where the nonabsorbing layer has the same refractive index as the sample. The absorption peak is centered at 1000 cm^{-1} (wavelength 10 μm).

nentially. The evanescent wave decays much faster in air than it does in a nonabsorbing medium with the sample matched refractive index, in full agreement with our expectations. However, it is instructive to compare these curves with a straightforward calculation of the penetration depth (Eq. 5.4). Keep in mind that the expression in Equation 5.4 gives us the depth at which the electric field of evanescent wave decays to 1/e of its surface value. The absorption of light is proportional to the wave intensity, which in turn is proportional to the square of the field. Since the field decays exponentially, the intensity decays twice as fast as the field. Thus, we would expect the absorbance to fall to 1/e of its value at the distance of $d_p/2$ into the sample.

The formula (Eq. 5.4) gives for the case of air $d_p = 1.15\ \mu m$, and for the nonabsorbing layer of the matching refractive index, $d_p = 1.85\ \mu m$. Taking one-half of the above values does not bring us to a precise agreement with the curves in Figure 11.11, which fall to 1/e of the surface value for the thickness corresponding to the absorbance value of just under 0.2. For the case of air gap, this occurs near the thickness of 0.3 μm, roughly at half of the value predicted by Equation 5.4.

For the second case, the agreement is much better, as it would be expected since the index matching at the second interface minimizes the effect of the interface. This simple example shows both the strengths and the limitations of the depth penetration concept. If a sample is highly inhomogeneous, the model most likely is not going to be very accurate.

The above case with a nonabsorbing layer of matching refractive index sandwiched between the sample and ATR element also confirms our intuitive picture of an evanescent wave that travels along the ATR interface with the portion of the wave that travels through the nonabsorbing medium not absorbed along the way, while the portion of the wave that travels through the absorbing medium is absorbed along the way. Hence, the absorbance in this case is the absorbance that would occur for the zero thickness of the nonabsorbing layer multiplied by the fraction of the total wave intensity that travels through the absorbing layer.

In the case when the nonabsorbing layer is air, the second interface renormalizes the strength of the electric field just behind the interface so the fraction of the intensity traveling through the absorbing medium is no longer controlled by the simple exponential decay that gives rise to the concept of the penetration depth. There is now a discontinuity in the evanescent wave intensity occurring at the second interface which can significantly modify the fraction of light traveling beyond the second interface.

11.10 REFLECTANCE AND TRANSMITTANCE OF A MULTIPLE INTERFACE SYSTEM

It is possible to derive expressions for the reflectance and the transmittance for a system consisting of any number of plane parallel interfaces separating different optical media. In the medium of incidence, incident and reflected light are traveling in different directions. Between any two interfaces, there are also two components of light, one traveling away from the first interface and one traveling toward it. The situation is schematically depicted in Figure 11.12.

Incident light with amplitude A_0 impinges at interface 1, partially reflects and partially transmits through interface 1, propagates through medium 1 to interface 2, partially reflects and partially transmits through interface 2, and so on. Figure 11.12 shows light propagating perpendicular to the interfaces, but in what follows, we consider a more general case of light of a given polarization (either p or s) incident at the first interface with the angle of incidence θ.

Each component reflects at every interface that it encounters, and the reflected component then contributes to the component propagating in the opposite direction. Let us refer to the incident medium as medium 0. The last (transmission) medium is the one into which light transmits after traversing all the interfaces. If there are N interfaces, then the transmission medium is the Nth medium. Let us also choose the axis perpendicular to the interfaces as the z-axis, with the positive direction being the direction from the incident toward the transmission medium. The incident medium contains two components of light: incident and reflected. The transmission medium, however, contains only transmitted light.

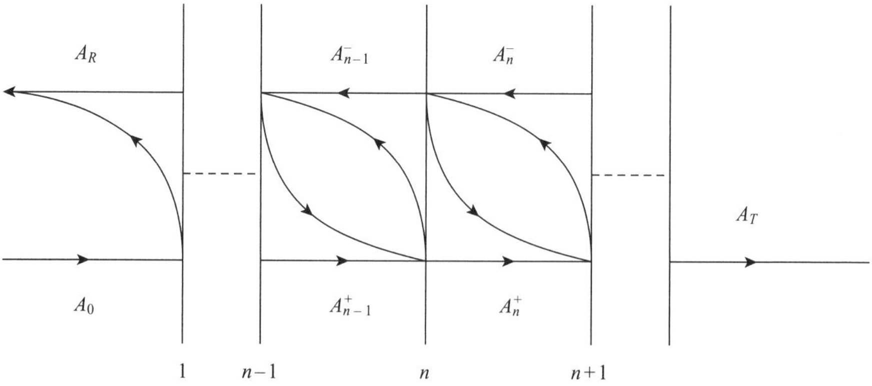

Figure 11.12 Propagation of light through multiple parallel interfaces.

Figure 11.12 shows the first and last interfaces and a general nth interface together with the one preceding it, $(n-1)$, and the one following it, $(n+1)$. Note that, in this case, it would be next to impossible to do what we did for the simple two-interface system, that is, to follow each component as it reflects and transmits through the various interfaces and try to add them up. So what we are using here is a sort of a trick. We are not following a particular component; we simply observe that the reflections at all these interfaces will generate an infinite number of components and that all these components will sort themselves into two groups, one advancing in the positive direction of the z-axis and one in the negative direction. Furthermore, these components all have the same frequency since they all descend from the same incident wave. If we add together all the multiple components advancing in the positive direction, we get A_+. Similarly, if we add all those advancing in the opposite direction, we get A_-. Next, we observe that the Fresnel reflection and transmission coefficients at a particular interface govern each of the components in the infinite sums A_+ and A_-, so therefore, they apply to the components A_+ and A_- themselves. Thus, we can connect the components A_+ and A_- on two sides of an interface by the same Fresnel amplitude coefficients that we use to connect incident transmitted and reflected components of a single plane wave across the interface. Therefore, for the nth interface, we can write

$$A_n^+(0) = t_{n-1,n}A_{n-1}^+(d_{n-1}) + r_{n,n-1}A_n^-(0) \tag{11.15a}$$

$$A_{n-1}^-(d_{n-1}) = t_{n,n-1}A_n^-(0) + r_{n-1,n}A_{n-1}^+(d_{n-1}). \tag{11.15b}$$

Note the values in parentheses. They refer to the distance from the left interface of the layer. At the left interface of the layer, this distance is zero, and at the right interface, it is equal to the thickness of the layer. The x-y coordinates are the same for all the components and are not explicitly indicated.

Polarization is also not explicitly indicated but is assumed to be either s or p. As we have said, a wave propagates within a layer as a plane wave. So, we can write the expression for a component at the right interface of a layer in terms of its value at the left interface as

$$A_n^{\pm}(d_n) = A_n^{\pm}(0)e^{\pm 2\pi i k n_n \cos(\varphi_n) d_n}. \tag{11.16}$$

The sign + refers to the forward, and the sign—to the backward propagating component. n_n is a generally complex refractive index of the nth layer, k is the wave number of the wave in vacuum, and φ_n is the angle of refraction into the nth layer. The ± sign in the exponent of the propagation factor refers to the direction of propagation with respect to the z-axis. Snell's law states that

$$n_1 \sin(\varphi_1) = \cdots = n_n \sin(\varphi_n) = \cdots. \tag{11.17}$$

According to Equation 11.17, the angle φ_n can be expressed in terms of the angle of incidence θ at the first interface as follows:

$$n_n \cos(\varphi_n) = \sqrt{n_n^2 - n_1^2 \sin^2(\theta)}, \tag{11.18}$$

which further simplifies the expression (Eq. 11.16). The system of two Equation 11.15 can be solved in terms of the fields in the $(n-1)$th layer:

$$\begin{aligned} A_n^+(0) &= \frac{t_{n-1,n}t_{n,n-1} - r_{n,n-1}r_{n-1,n}}{t_{n,n-1}} A_{n-1}^+(d_{n-1}) + \frac{r_{n-1,n}}{t_{n,n-1}} A_{n-1}^-(d_{n-1}) \\ A_n^-(0) &= -\frac{r_{n-1,n}}{t_{n,n-1}} A_{n-1}^+(d_{n-1}) + \frac{1}{t_{n,n-1}} A_{n-1}^-(d_{n-1}). \end{aligned} \tag{11.19}$$

We have put the Equation 11.19 into a form suggestive of matrix manipulations. We can define a two dimensional vector:

$$A_n(0) = \begin{pmatrix} A_n^+(0) \\ A_n^-(0) \end{pmatrix}, \tag{11.20}$$

so we can simplify Equation 11.19 to

$$\begin{pmatrix} A_n^+(0) \\ A_n^-(0) \end{pmatrix} = \frac{1}{t_{n,n-1}} \begin{pmatrix} 1 & r_{n,n-1} \\ -r_{n-1,n} & 1 \end{pmatrix} \begin{pmatrix} A_{n-1}^+(d_{n-1}) \\ A_{n-1}^-(d_{n-1}) \end{pmatrix}. \tag{11.21}$$

We have taken into account that

$$t_{n-1,n}t_{n,n-1} - r_{n,n-1}r_{n-1,n} = 1. \tag{11.22}$$

The correctness of the Equation 11.22 is easy to verify by direct calculation. We can now also recast the expression (Eq. 11.16) into the matrix form

$$\begin{pmatrix} A_n^+(d_n) \\ A_n^-(d_n) \end{pmatrix} = \begin{pmatrix} e^{2\pi i k d_n \sqrt{n_n^2 - n_0^2 \sin^2(\theta)}} & 0 \\ 0 & e^{-2\pi i k d_n \sqrt{n_n^2 - n_0^2 \sin^2(\theta)}} \end{pmatrix} \begin{pmatrix} A_n^+(0) \\ A_n^-(0) \end{pmatrix}. \tag{11.23}$$

Equations 11.21 and 11.23 allow us to express $A_n(0)$ in terms of $A_{n-1}(0)$. The incident and reflected waves form the first vector:

$$A_1 = \begin{pmatrix} A_{in} \\ A_r \end{pmatrix}. \tag{11.24}$$

The transmitted wave is the sole component of the last (Nth) vector:

$$A_N = \begin{pmatrix} A_t \\ 0 \end{pmatrix}. \tag{11.25}$$

This enables us to use Equation 11.23 as a recursive relation and to cast it into the form in which the transmitted wave is expressed in terms of the incident and reflected waves. Thus, we get

$$\begin{pmatrix} A_t \\ 0 \end{pmatrix} = \left(\prod_{n=2}^{N} \frac{1}{t_{n+1,n}} \begin{pmatrix} 1 & r_{n+1,n} \\ -r_{n,n+1} & 1 \end{pmatrix} \begin{pmatrix} e^{2\pi i k d_n \sqrt{n_n^2 - n_1^2 \sin^2(\theta)}} & 0 \\ 0 & e^{-2\pi i k d_n \sqrt{n_n^2 - n_1^2 \sin^2(\theta)}} \end{pmatrix} \right)$$
$$\frac{1}{t_{21}} \begin{pmatrix} 1 & r_{21} \\ -r_{12} & 1 \end{pmatrix} \begin{pmatrix} A_{in} \\ A_r \end{pmatrix}. \tag{11.26}$$

Note the structure of the expression (Eq. 11.26). There are two types of matrices in the expression: the interface type shown in Equation 11.21 and the propagation type shown in (Eq. 11.23). The interface matrix connects the fields of the two oppositely traveling waves on two sides of an interface. It has a suggestive structure. The off-diagonal elements connect the components that travel in the opposite directions, while the diagonal elements connect the components that travel in the same direction. The matrix shown in Equation 11.23 describes the propagation of the two components through a layer from one interface to the next. This matrix is diagonal. It contains the propagation factors for the two oppositely traveling waves. The minus sign in the exponent of the exponential function for the lower component is a consequence of this wave traveling in the opposite direction to the incident wave.

The expression (Eq. 11.26) than can be understood as a sequence of matrices acting in succession, each describing either propagation through an interface or propagation through a layer. The matrix product of all the matrices in the parentheses of the expression (Eq. 11.26) results in a 2×2 matrix M. We can proceed by writing Equation 11.26 as

$$\begin{pmatrix} A_t \\ 0 \end{pmatrix} = \begin{pmatrix} M_{11} & M_{12} \\ M_{21} & M_{22} \end{pmatrix} \begin{pmatrix} A_{in} \\ A_r \end{pmatrix}. \tag{11.27}$$

This immediately leads to the solution for the transmittance and reflectance amplitude coefficients of the system consisting of any number of parallel interfaces:

$$\rho = \frac{A_r}{A_{in}} = -\frac{M_{21}}{M_{22}} \tag{11.28a}$$

$$\tau = \frac{A_t}{A_{in}} = M_{11} - \frac{M_{12}M_{21}}{M_{22}}. \tag{11.28b}$$

The above results are the generalization of the single interface Fresnel amplitude coefficients. We have separately derived the two-interface amplitude coefficients (Eq. 11.6) and have investigated some of the complexities hidden in these seemingly simple expressions. The results (Eq. 11.6) that we separately derived earlier also follow from Equations 11.26 and 11.28 by specifying that the number of interfaces is two.

The expressions (Eq. 11.28) are of a great practical interest to the designers of optical coatings. Often, a large number of thin layers are coated over an optical surface either to enhance the reflectance of the surface or to diminish it. Another use of multiple optical coatings is to restrict the transmittance (or the reflectance) of an optical surface to a particular spectral range. Using computers to model these multilayer coatings, designer optical coatings, having almost arbitrary transmittance/reflectance characteristics, can be developed.

Note that unlike the Fresnel amplitude coefficients that described propagation through a single interface, the amplitude coefficients for the multiple interface systems contain propagation factors that, as we have seen earlier, carry information about light absorption in the material. This prevents the generalized amplitude coefficients from obeying the type of relationships such as Equation 11.22.

Another phenomenon that can be analyzed using the general expression for multiple interfaces (Eq. 11.26) is the propagation of light through a medium in which the optical constants vary continuously along one direction. An example is a material that has been modified near the surface by the diffusion of another material into it. For instance, water could diffuse into glass. The very surface layer of glass has the largest density of water molecules, and this density quickly decays deeper into the bulk. The optical constants of glass are modified by the presence of water molecules and these "constants" change continuously with the distance from the surface until, at some depth, they asymptotically reach the values for pure glass. A description of such a surface could utilize a multilayer model with a large number of layers with optical constants changing in very small steps from one layer to the next. This model could be used to help interpret ATR depth profiling data where a continuous change in optical constants near the surface of a material is anticipated.

12 Metal Optics

12.1 ELECTROMAGNETIC FIELDS IN METALS

Metals comprise an important class of optical materials. Metals are almost universally used as the material for making mirrors. A metal could be used as a polished piece of solid metal or evaporated as a film over a polished surface of some not particularly reflective material such as glass. Except in a few special cases, metal is used as an opaque material that reflects light and does not transmit it. A very thin metal film, just a few hundred atomic layers thick, is fully opaque to light.

It is worth briefly recalling why metals are so highly reflective. Metals contain electrons moving freely in a background of positive ions. Electrons are confined to metal and cannot escape from the surface of the metal because they are electrically bound to the positive ions in the metal. As an electric field is applied across a metal slab, freely moving electrons are pushed by the electric force to one surface of the metal, leaving the opposite surface positively charged. These charges, induced by the applied field, generate surface charge densities that create inside the metal slab an electric field opposite to the externally applied field, thus weakening the total field inside the metal. As long as the field inside the metal persists, it drives more free electrons to the surface, thus increasing the surface charge density and hence increasing the strength of the induced opposing field, thus reducing the total electric field inside the metal. This process stops when the electric field induced by the surface charge density equals the applied external field, therefore eliminating the overall electric field in the metal. A zero electric field in the metal means there is no force on the remaining free electrons, so the process stops.

Since electrons have extremely small masses, the entire balancing process occurs extremely fast so as the electric field of an electromagnetic wave reaches the metal surface, the electrons quickly rearrange themselves on the surface to generate the opposing field of equal

Internal Reflection and ATR Spectroscopy, First Edition. Milan Milosevic.

strength, thus cancelling the total field inside the metal. It follows that electromagnetic waves cannot propagate through metal. The electrons oscillating near the surface of the metal in effect generate an electromagnetic wave inside the metal that completely cancels the incoming wave inside the metal, while outside of the metal we see it as the reflected wave.

These motions of free electrons inside the metal to quickly redistribute themselves in order to generate an opposing field, which, inside the metal, cancels the incoming electric field, are what is responsible for the high optical reflectivity of metals. Also, this explains why better electrical conductors are generally better reflectors.

However, keep in mind that the tendency of electrons to rearrange their position in metal to cancel the incoming electric field so that no net electric field can exist in metals is not some mysterious electronic conspiracy; it is, as we have seen, just electrons being pushed by the net electric field until the net field becomes zero (Fig. 12.1).

It is possible to extend the harmonic oscillator model that we used to describe the optical properties of dielectric materials to the description of metals. All we have to recognize is that the existence of the free moving electrons clearly implies that no restoring harmonic force acts on them. Thus, in Equation 4.10, at least one term in the sum has an associated normal mode frequency equal to zero.

As electrons in the metal rearrange themselves to cancel the external field inside the metal, the constituent atoms inside the metal experience

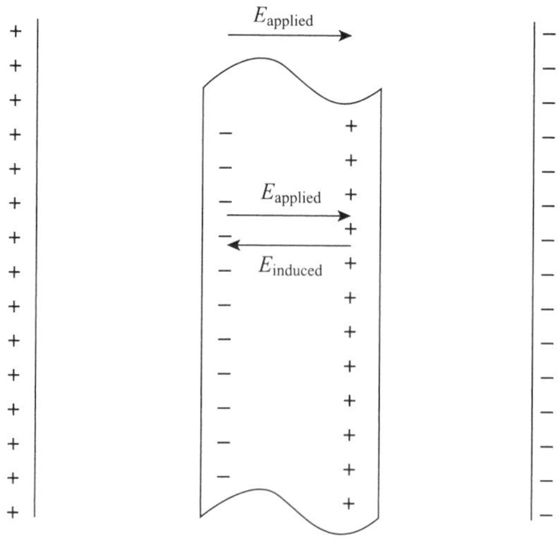

Figure 12.1 A metal slab inside a capacitor showing the surface charges and associated electric fields.

a zero net field so the induced polarization of atoms vanishes as well. The result is that only the free electron term in the sum survives:

$$\alpha(\omega) = \frac{e^2}{m}\frac{f_0}{-\omega^2 + i\gamma_0\omega}, \tag{12.1}$$

where m is the mass of electron.

This is an interesting result. It states that the polarizability of metals is infinite for static fields ($\omega = 0$). Based on the above discussion, this is not too surprising. The dielectric constant is then simply

$$\varepsilon(\omega) = 1 + 4\pi N\alpha(\omega) = 1 + i\frac{4\pi Ne^2 f_0}{m\omega(\gamma_0 - i\omega)}, \tag{12.2}$$

where we could use the direct relationship between polarizability and dielectric constant instead of the Clausius–Mossotti equation (Eq. 2.11) because there is no net field in the metals, no polarization of atoms, and hence no polarization-induced interaction between atoms. Since the Clausius–Mossotti equation was developed to include the polarization-induced interactions between atoms, it does not apply to these circumstances.

Let us now return to the last Maxwell equation in Equation 1.3. The equation was valid for a space without charges and currents. If currents are present, the equation changes into

$$\nabla \times \boldsymbol{B} = \frac{4\pi}{c}\boldsymbol{j} + \frac{1}{c}\frac{d\boldsymbol{E}}{dt}. \tag{12.3}$$

For a harmonic electric field, $\boldsymbol{E}(t) = \boldsymbol{E}_0 e^{-i\omega t}$, the time derivative in Equation 12.3 becomes

$$\frac{d\boldsymbol{E}}{dt} = -i\omega\boldsymbol{E}. \tag{12.4}$$

According to Ohm's law, the electric field $\boldsymbol{E}$ induces in metals a current of density $\boldsymbol{j}$, and the conductivity is defined as a constant of proportionality between the current and the field:

$$\boldsymbol{j} = \sigma\boldsymbol{E}. \tag{12.5}$$

Inserting Equation 12.5 into Equation 12.4 yields

$$\nabla \times \boldsymbol{B} = \left(\frac{4\pi}{c}\sigma + \frac{1}{c} i\omega \right) \boldsymbol{E} = -i\frac{\omega}{c}\left(1 + i\frac{4\pi\sigma}{\omega}\right) \boldsymbol{E}. \tag{12.6}$$

The term in the parenthesis of the right-hand side of Equation 12.6 is a constant, so we could use Equation 12.4 again to express Equation 12.6 as

$$\nabla \times \boldsymbol{B} = \left(1 + i\frac{4\pi\sigma}{\omega}\right)\left(-i\frac{\omega}{c}\boldsymbol{E}\right) = \left(1 + i\frac{4\pi\sigma}{\omega}\right)\frac{1}{c}\frac{d\boldsymbol{E}}{dt} = \frac{1}{c}\frac{d}{dt} \in(\omega)\boldsymbol{E}, \tag{12.7}$$

where, instead of using Ohm's law to explicitly account for currents induced by the applied field, we assigned the entire response of the medium to the applied electric field to the dielectric constant of the medium. This may seem as a somewhat artificial assignment, but it is somewhat arbitrary how we should divide the response of the medium to the applied external field between the dielectric constant and conductivity. In a nonconductive medium, we would set $\boldsymbol{j} = 0$ on the right-hand side of Equation 12.3 and incorporate the effects of the medium by multiplying the electric field with the dielectric constant of the medium. We can therefore try to generalize the concept of dielectric constant to incorporate the induced currents. It thus seems worthwhile to pursue this line of reasoning to the end and see where it leads.

By comparing Equation 12.7 with Equation 12.6, and by using the result (Eq. 12.2), we arrive at

$$\sigma = \frac{Ne^2 f_0}{m(\gamma_0 - i\omega)}. \tag{12.8}$$

The result (Eq. 12.8) is the same as given by the Drude model for the electrical conductivity of metals in the static limit ($\omega = 0$) in turn justifying the generalization above. This result shows that the formalism developed for the electromagnetic treatment of optical materials such as glass can be extended to a totally different class of materials such as metals. The response of a material to an incoming electromagnetic field that yields a description of the reflection and the transmission of electromagnetic waves by optically transparent materials can be stretched to the point that it describes the electrical conductivity of metals. This demonstrates the extraordinary robustness of the electromagnetic theory.

It may be of some interest to pause for a moment and to look back over how the original electromagnetic theory of Maxwell was gradually stretched to cover more than it was originally developed to describe. First, as discovered by Maxwell himself, the theory predicted the existence of electromagnetic waves that propagate through empty space with the speed of light. The identification of these waves with light represented a great conceptual unification in our understanding of nature, the unification of electric, magnetic, and optical phenomena, all under a single formalism. The formalism then correctly accounted for the reflection of light at a surface of a transparent medium, the propagation of light through a medium, the absorption of light by a medium, and so on. As the formalism was pushed further to describe internal reflection, and in particular supercritical internal reflection, it performed flawlessly. Most of what we have done in the past chapters was to follow this formalism and to explore the various aspects of it. The formalism correctly described the evanescent wave in the supercritical regime of the internal reflection and accounted for all its various properties. It is fascinating that the formalism is able to incorporate absorbing materials, and it comes as a pleasant surprise that the formalism can be stretched so far to cover not only the optical properties of metals but also the electrical properties as well.

We have to think back about the concept of refractive index to see how much of a conceptual stretch it is to apply it to metals. The original concept of refractive index was developed around Snell's law of refraction. The explanation of Snell's law, and hence of the refractive index, was based on the assumption that the speed of light is different in different optical media. The ratio of the speed of light in vacuum to the speed of light in a medium was defined as the refractive index of the medium. That explained Snell's law.

Later on, the development of Maxwell's electromagnetic theory connected the refractive index to the dielectric constant of the material. Further on, the concept of the refractive index was extended to light-absorbing materials by allowing the refractive index to become complex, wherein the imaginary part of the complex refractive index was responsible for light absorption.

We have seen how Fresnel formulae easily incorporated this extension of the concept of refractive index to absorbing media and, at the same time, how the model of polarizability, based on the atomic and molecular structure of materials, naturally incorporated the same extension. And now this concept of refractive index is extended to metals, the materials that contain freely moving charges and thus actively resist any electric field to be established within the metal. However, as we can see, everything fits well together and provides a consistent framework for the description of electromagnetic waves.

12.2 PLASMA

Going back to Equation 12.2, an optically interesting limit is that of high frequency. In the high frequency limit, $\omega \gg \gamma_0$, so Equation 12.2 becomes

$$\varepsilon(\omega) = 1 - \frac{4\pi Ne^2 f_0}{m\omega^2}. \tag{12.9}$$

We can define

$$\omega_p^2 = \frac{4\pi Ne^2 f_0}{m}. \tag{12.10}$$

The frequency ω_p is called plasma frequency. Plasma is the state of mater where electrons are stripped from atoms, leaving an electrically neutral mix of positively charged ions and free electrons. Such a state of mater can be caused by high temperature, such as exists in stars, or it can be caused by electomagnetic fields in very low density gases. Metals also show some properties of plasma with the positive ions in metals locked into the crystal lattice. By inserting values typical for metals into Equation 12.10, one finds that the plasma frequency of metals falls roughly into the UV region of the spectrum.

With the definition (Eq. 12.10), Equation 12.9 takes a particularly simple form:

$$\varepsilon(\omega) = 1 - \frac{\omega_p^2}{\omega^2}. \tag{12.11}$$

We see immediately that the plasma frequency ω_p marks the transition point at which the dielectric constant of a metal switches from negative at low frequencies (below ω_p) to positive at high frequencies (above ω_p). Since the refractive index is the square root of the dielectric constant, the refractive index switches from purely imaginary, below ω_p, to real, above ω_p. Thus, for frequencies below ω_p, the metal is opaque to light and at frequencies above ω_p, metals turn transparent. This is the well-known phenomenon of UV transparency of metals.

The same analysis applies to the ionosphere, a layer of the upper earth atmosphere that is kept in the state of partial plasma by the constant bombardment from extraterrestrial radiation. That is why we can hear radio stations from around the world on long radio waves. Long waves reflect from the ionosphere and short waves just transmit through and get lost in space.

The phenomenon of UV transparency of metals can be used to experimentally measure ω_p for different metals by measuring the reflectance of a metal as a function of the wavelength. The reflectance should be high right up to the wavelength of light λ_p that corresponds to the frequency ω_p. After that, it should suddenly drop to near zero and then slowly continue to rise as the frequency increases (wavelength decreases). These measurements have been done and a remarkable agreement with the values predicted by Equation 12.10 was observed. For instance, for lithium, the predicted and measured λ_p are identical at 155 nm. For sodium, the predicted λ_p is 209 nm, while the measured λ_p is 210 nm. The agreement is within 10% for most metals, thereby further encouraging the extension of the above formalism to metals. Therefore, it seems justified to treat a metal as just another optical medium as long as we provide the appropriate optical constants for it.

12.3 REFLECTANCE OF METAL SURFACES

Figure 12.2 shows the reflectance of a metal with refractive index $n = 10 + 20i$ versus angle of incidence for both polarizations. Note that the shapes of the curves exhibit the general characteristics of external reflection. There is a Brewster's angle for p polarization and the reflectance of both polarizations becomes total at 90°. This is not surprising given the large real part of the refractive index. That is also why the

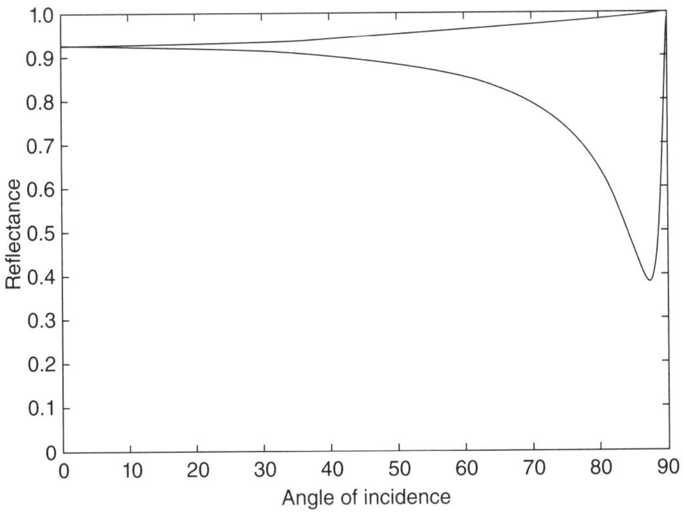

Figure 12.2 The reflectance of a metal ($n = 10 + 20i$) surface in air for parallel (lower trace) and perpendicular polarizations. Note the presence of the Brewster angle revealed by the dip in the reflectance for parallel polarization near 87°.

Brewster's angle is so high. The reflectivity displayed in Figure 12.2 is above 90%, which is typical for metals.

Let us now imagine that a metal film is placed as a coating over a surface of a high refractive index material such as ZnSe with refractive index $n = 2.4$ and that light is incident onto the interface from the ZnSe side. At first sight, that would look like internal reflection, but note that the real part of the refractive index of metal is much larger than the refractive index of ZnSe. That implies that the reflection at the interface is still external.

The clarity of the situation is muddied by the very high value of the imaginary part of the refractive index of metal. So the simplest way to assess what type of reflection this is is to simulate it. The result is shown in Figure 12.3. Again, the overall shapes of the reflectance curves indicate external reflection although light is incident from a high refractive index material.

The reflectivity of the ZnSe–metal interface is lower than the reflectivity of the air–metal interface. Also, Brewster's angle is lower as well. This is consistent with the relative refractive index of the air–metal interface being more than twice as large as the relative refractive index of the ZnSe–metal interface.

Finally, the question may arise whether the reflection is external because the real part of the refractive index of metal was larger than

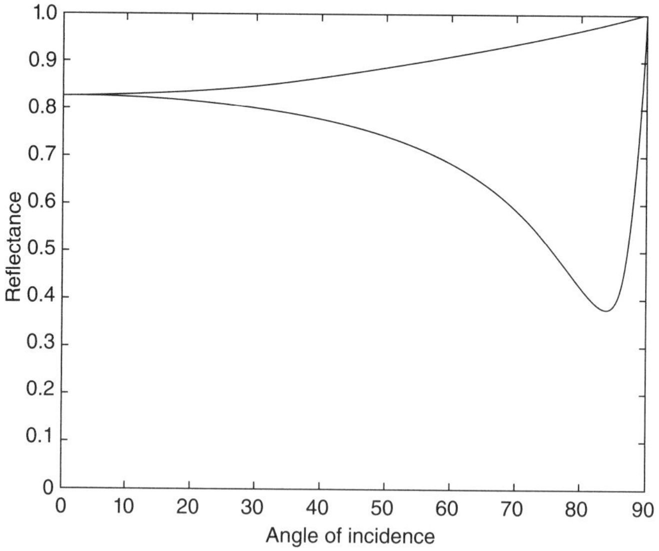

Figure 12.3 Reflectivity of a metal–ZnSe surface for light incident from ZnSe versus the angle of incidence. The reflectivity of the metal–ZnSe surface is significantly lower than that of metal–air surface.

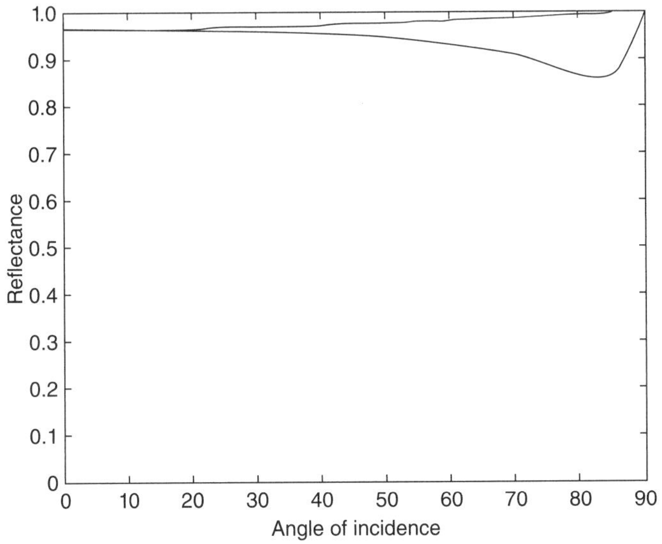

Figure 12.4 Reflectivity of a metal–ZnSe interface versus the angle of incidence. The refractive index of metal is $n = 1.5 + 20i$.

the refractive index of ZnSe or would it still be external if the real part of the refractive index of metal was smaller than that of ZnSe?

This is easily answered by changing the refractive index of metal to $n = 1.5 + 20i$. We are not here concerned whether a metal with such a refractive index exists; all we want to know is if that would change the nature of the reflection. Figure 12.4 shows the result. Clearly, the reflection is still external. The presence of the overwhelming imaginary part of the refractive index of the metal continues to keep the nature of the reflection external-like.

The metal in these simulations was thick enough so that, for all intents and purposes, the thickness of the metal was semi-infinite. In practice, a thin metal layer just a few micrometers thick is effectively semi-infinite. We will investigate thin metal films below. But let us first discuss why metals are so effective reflectors of light. Note that it is not so because of the large real part of the refractive index of metals. As we have seen above, a metal with the real part of the refractive index as small as 1.5 has produced a reflectance above 80%.

We can easily calculate what the reflectivity would be for a material that has a large refractive index, say, $n = 10$, but has a zero absorption index. The result is that the reflectivity is very much lower than the 90% typical of metals. Thus, the real reason for the high reflectivity of metals must be their large absorption index. This is somewhat counterintuitive. The absorption index of a medium is responsible for light absorption by the medium. High reflectivity means the material reflects light and does not absorb it. The take-home message of the above analysis is that only the fraction of incident light that is not reflected by the surface of a material can enter into the material and become absorbed by it. The very high absorption index of metals very effectively prevents light from entering metal, and this is why so little of the incident light is absorbed by metals. Light simply cannot get in. However, whatever gets in is absorbed very quickly.

The nature of absorption of light in metal, as we have seen, is essentially due to the electrical resistance in the metal. The electric field of the incident light produces electric currents in the surface of the metal. These electric currents, oscillating in the rhythm of the incident wave, generate the reflected wave. However, since electrons in metal cannot move completely freely due to electrical resistance, the reflected wave is slightly less intense than the incoming wave. The reflected wave is not really reflected in the same sense that, for example, a ball bounces off the wall. The mechanism is as follows. The electric currents induced in the surface of the metal by the incident field are oscillating at the same frequency as the incident wave. Oscillating currents produce new electromagnetic waves. It is easy to see how this mechanism produces a reflected wave that obeys the law of reflection. Inside the metal, the two waves interfere destructively, thus in effect preventing the electromagnetic field from entering. Therefore, the electrical conductivity of metals is responsible for the high reflectivity of metals, while the resistance of metals (which is, by definition, the inverse of electrical conductivity) is responsible for the absorption of light in the metal.

12.4 THIN METAL FILMS ON TRANSPARENT SUBSTRATES

Up to now, we have studied the reflectivity of an interface between a semi-infinite metal medium and a transparent optical medium. Now, we turn to the reflective properties of thin metal films coated over the surface of a transparent optical medium.

As we will see, in the regime of internal reflection, these films exhibit quite an extraordinary behavior.

To illustrate, let us start with the internal reflectance of a thin metal film (refractive index $n = 10 + 20i$). Light of wavelength 10 μm is incident from a nonabsorbing optical material (such as glass) having a refractive index of $n = 1.5$ at an angle of incidence of 45°. Behind the metal film is air (or vacuum). Figure 12.5 shows the reflectance for s and p polarizations of light for a thin metal film, as specified above, versus the thickness of the film.

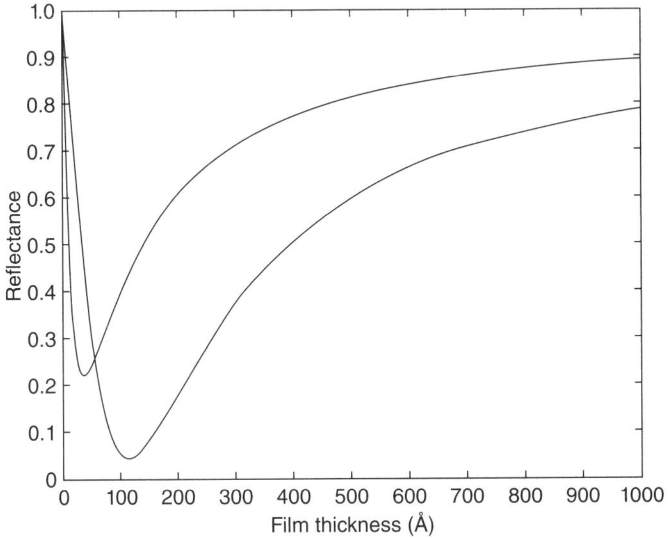

Figure 12.5 Internal reflectance of a thin metal film versus film thickness. The upper curve is for the *s*-polarized and the lower curve for *p*-polarized incident light. The angle of incidence is 45°.

The important observation is that the angle of incidence is supercritical and that, in the absence of the metal film, the reflectivity of the interface would be total. That is why the reflectance of both polarizations starts at one (i.e., total) for zero thickness of the metal film. What is surprising is the dip that reflectance for both polarizations takes for very thin films. The dip in the reflectance could occur for two reasons. Either light is transmitted into the next medium or it is absorbed by the film. In this case, light cannot transmit into the next medium behind the film (air or vacuum) since the angle of incidence is supercritical. Therefore, light is absorbed by the film. We see that the reflectivity of the two polarizations returns to expected levels for thicker films (1000 Å = 0.1 μm). However, the dips are puzzling. They occur for both polarizations but are maximized at different thicknesses. Note also that the reflectance of *s*-polarized incident light is higher than the reflectance of *p*-polarized light for extremely thin films and then reverts to the usual behavior for thicker films.

An interesting question is if one can choose the angle of incidence and the film thickness for which the reflectance of *p*-polarized incident light would become zero. It turns out that choosing the angle of incidence of 50° rather than 45° makes the reflectance for *p*-polarized incident light essentially zero for the film thickness of $d = 94$ Å (Fig. 12.6).

The actual details of course depend on the particular values of refractive indices for the materials involved. However, it is surprising that a

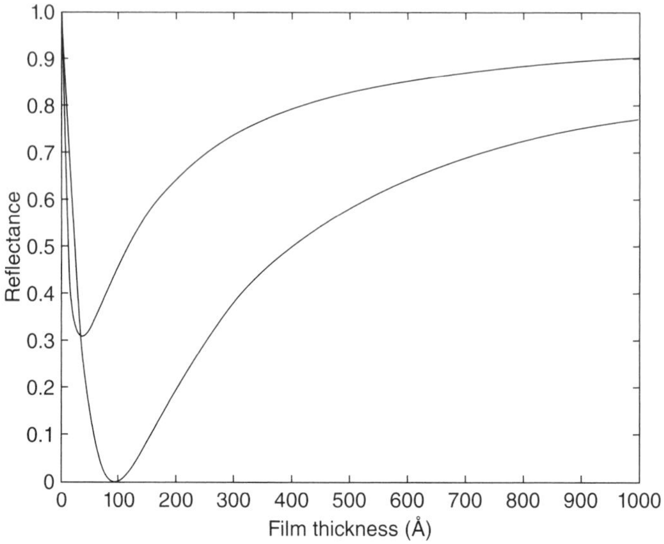

Figure 12.6 Internal reflectance of a thin metal film versus film thickness. The upper curve is for the *s*-polarized and the lower curve for *p*-polarized incident light. The angle of incidence is 50°.

very thin film of metal could act as a highly effective antireflection coating! The *p*-polarized incident light at the given film thickness is totally absorbed by the film. Nothing is reflected!

Another surprise coming from thin metal films is shown in Figure 12.7. An extremely thin film (30 Å) is coated over an optically transparent medium with the refractive index $n = 1.5$. Light is incident from within the medium. Behind the metal film is air (or vacuum). The reflectivity for both polarizations for subcritical angles of incidence is as would be expected. It is well-known that very thin metal films are semitransparent. They are regularly used as beam splitter coatings where a fraction of incident light is transmitted through the film and a fraction is reflected. So for a subcritical angle of incidence (in our case below about 42°), some of the incident light that is not reflected is transmitted through and a fraction is absorbed by the film.

12.5 CURIOUS REFLECTANCE OF EXTREMELY THIN METAL FILMS

However, the surprising behavior occurs at supercritical angles of incidence. First, right away it is surprising that the reflectance for the

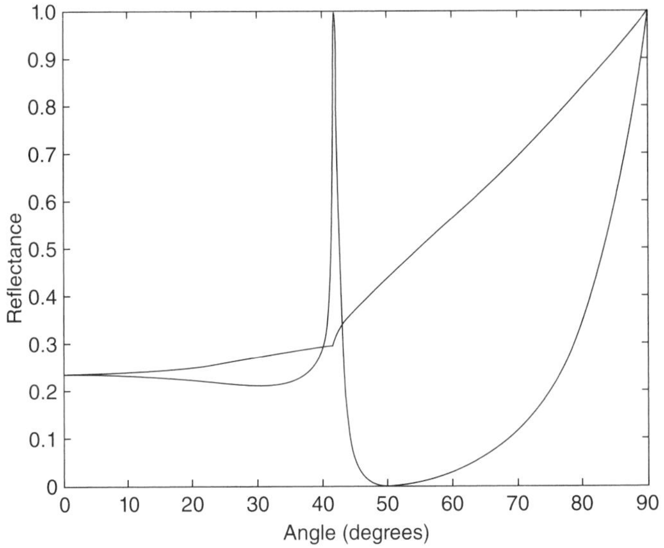

Figure 12.7 Reflectance of a 30-Å thick metal film versus the angle of incidence for *p*-polarized (graph with sharp spike) and *s*-polarized incident light.

p-polarized light shoots up to 100% exactly at the critical angle. Note that in Figure 12.7, aside from a minimal wiggle at the critical angle, the reflectance for the *s*-polarized light proceeds without dramatic upheavals from normal to grazing incidence.

The film is extremely thin, but it still manages to absorb most of the incident light of both polarizations at supercritical angles of incidence. Note that, for supercritical incidence, there is no transmitted light. The entire fraction of incident light that was not reflected was absorbed by the film.

For *p*-polarized light, the reflectance of the film drops to zero near 50°. Obviously, there must be a mechanism underlying this near total absorption. It turns out that this mechanism is related to the so-called surface plasma waves (plasmons) excited not on the attenuated total reflection (ATR) material–metal interface but on the metal–air interface. The electromagnetic wave must travel through the entire film thickness to reach the second surface. As we have seen, electromagnetic waves have a hard time traveling through metal, but if the metal film is thin enough, a significant portion of the wave reaches the back surface of the film and excites surface plasma waves in the interface that absorbs a significant fraction of energy from the electromagnetic wave.

Surface plasma waves are oscillations of the surface charge density. Under normal circumstances, electromagnetic waves cannot excite surface plasma waves since plasmons propagate at different speeds than light waves. We will come back to the topic of surface plasmons later on to study this phenomenon in more detail.

The surprising drop of reflectance almost to zero due to absorption by the surface plasmons follows an equally surprising peak in reflectance to total (i.e., 100%) exactly at the critical angle. The lack of absorption of incident light by the otherwise prodigiously absorbing thin metal film is due to the fact that *p*-polarized light at the critical angle has its electric field perpendicular to the interface between the element and the metal film. The field perpendicular to the interface cannot drive electrons in the film along the interface. On the other hand, the film is too thin for a current to form perpendicular to the interface. If there are no currents in the metal film, then there cannot be resistive losses either. Since it is the resistive losses that would rob energy out of the incident wave, the reflectance is total. That is why the plunge of reflectance to zero over just several degrees up from the critical angle is so mystifying. Again, the fine points depend on the particular set of parameters used, but the general behavior persists for a reasonable range of variation in these parameters.

12.6 ATR SPECTROSCOPY THROUGH THIN METAL FILMS

It is sometimes of interest to cover an ATR element with a metal film. One might, for instance, be interested in the electrochemical properties of a compound and needs to subject it to an electric field while simultaneously spectroscopically observing the changes caused by the applied voltage. Or one may be interested in the catalytic effects of a metal surface. Studying these phenomena by ATR spectroscopy is a good way to do it since the evanescent wave penetrates only a short distance beyond the ATR crystal and the ATR signal is dominated by whatever is occurring near the surface of the ATR element.

However, for this to work, the evanescent field must penetrate through metal and extend to the sample behind the film. Obviously, the metal film has to be thin. We can model this situation by using the refractive index $n = 10 + 20i$ that we have seen gives a realistic description of metals and by simulating the sample by the same parameters that we used earlier to produce spectra shown in Figure 11.3.

The refractive index of the ATR material was selected to be 4 (Ge) and the angle of incidence 28.5°. The polarization is parallel (p). If we set the thickness of the metal layer to be 20-Å thick, the resulting spectrum is shown in Figure 12.8.

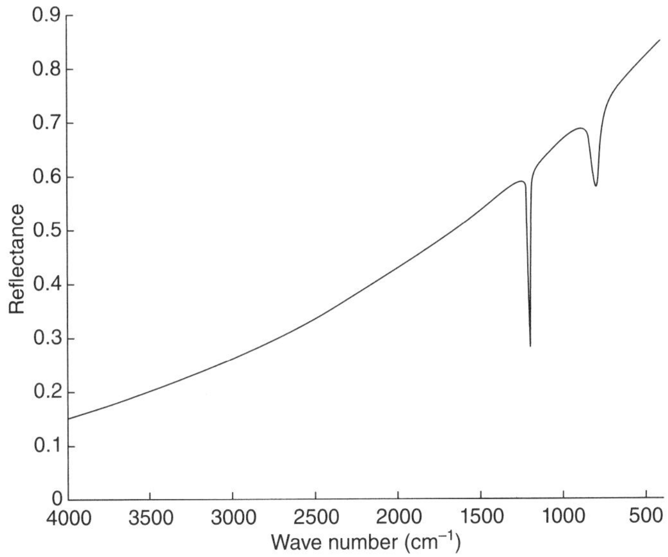

Figure 12.8 ATR spectrum of a sample as measured through a thin metal film coated over the ATR element.

The presence of absorption peaks in the spectrum clearly demonstrates that the evanescent wave is extended beyond the metal film. Note that the thickness of the film is only a few atomic layers, but the effect of the metal film on the spectrum is huge, reducing the transmittance to less than 20% at 4000 cm^{-1}. Nevertheless, the absorption peaks clearly come through demonstrating that such experiments can be conducted.

Note that in the above simulation, we assumed that the refractive index of metal is not changing with wavelength. This assumption is obviously not realistic. Our analysis of metals yielded Equation 12.11 for the dielectric constant, which showed a strong wavelength dependence of the dielectric constant and hence refractive index. However, the result shown in Figure 12.8 is still applicable since, as we have seen, it is important for the refractive index to be in the "metallic" range. Any value of the refractive index of a model metal within that range will produce the general behavior shown in Figure 12.8.

13 Grazing Angle ATR (GAATR) Spectroscopy

13.1 ATTENUATED TOTAL REFLECTION (ATR) SPECTROSCOPY OF THIN FILMS ON SILICON SUBSTRATES

An interesting special case of ATR spectroscopy is one of a very thin film deposited on a silicon substrate. By using a germanium ATR element and selecting the angle of incidence above 60°, the conditions for supercritical internal reflection for the Ge–Si interface are fulfilled regardless of the film thickness. The evanescent wave extending above the Ge ATR surface exists both within the thin film and within the silicon substrate.

The silicon substrate is nonabsorbing, so any absorption of the evanescent wave is due to the thin film, and even if this absorption is very weak due to the film's extremely small thickness, there is no spectral interference from the substrate, making it likely that the absorptions by the film are observable.

On first sight, this appears as just a good technique to analyze thin films on Si substrates. However, closer inspection shows that this is not just a good technique, that this is an exceptionally good technique. There is an enhancement phenomenon associated with this particular configuration that magnifies the absorption by a thin film under these circumstances by almost two orders of magnitude. This enhancement can be utilized to study extremely thin films—even monolayers.

The thickness of monolayers is about four orders of magnitude smaller than the thickness of typical films analyzed by infrared (IR) spectroscopy. Thus, the absorption levels associated with so extremely thin films are four orders of magnitude weaker than for typical samples. This makes these ultrathin films virtually unobservable.

Internal Reflection and ATR Spectroscopy, First Edition. Milan Milosevic.

13.2 ENHANCEMENT FOR *S*- AND *P*-POLARIZED LIGHT

It is important to understand the mechanism of the enhancement so that the resulting spectra could be properly interpreted. Let us start by illustrating the effect. We use a model material that has typical optical constants. A spectrum of a 10-Å-thick sample (for comparison, the diameter of hydrogen atom is ~1 Å) deposited on silicon using *p*-polarized light on a germanium ATR element for a 60° angle of incidence is shown in Figure 13.1 along with the spectrum of the same sample deposited directly onto the Ge ATR element, also for the *p*-polarization, but for the 24° angle of incidence—just above the critical angle for the sample–Ge interface. The difference is extraordinary. Closer inspection of the graphs reveals that the peaks of the spectrum recorded by grazing angle ATR (GAATR) are about 60 times stronger than the peaks recorded by regular ATR.

Keep in mind that the angle of incidence of 24° is rarely used in ATR spectroscopy and that a more typical angle of incidence of 45° would yield even weaker peaks. Thus, the film that would be on the edge of detectivity with regular ATR spectroscopy becomes a routine sample for GAATR spectroscopy.

Note that the standard ATR is generally considered a surface technique and it would be the technique best suited to analyzing thin films.

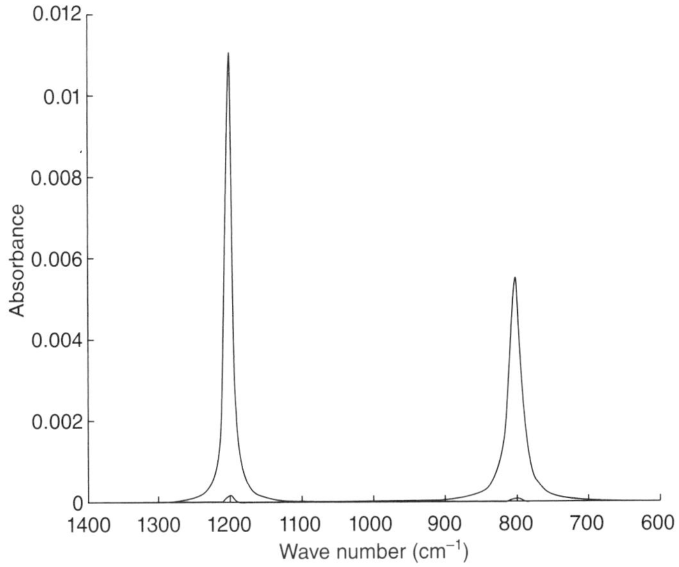

Figure 13.1 Comparison of the grazing angle ATR (upper curve) with regular ATR.

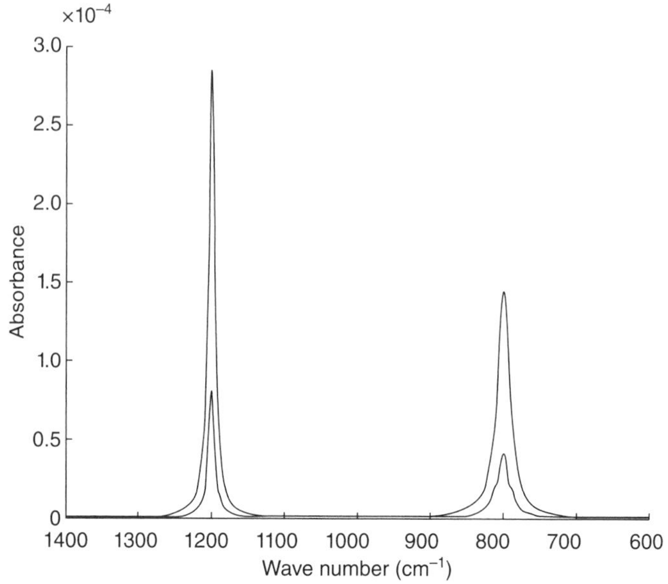

Figure 13.2 The comparison of absorption intensities for *s*-polarized light for a 10-Å-thick film on silicon substrate (upper curve) and the same film deposited directly onto the Ge ATR element for a 60° angle of incidence.

Silicon is a frequent choice of substrates for analyzing ultrathin films. It is hard, strong, and chemically inert and, thanks to the semiconductor industry, readily available in the form of wafers with highly polished and crystallographically oriented surfaces.

The absorption levels obtained by the *s*-polarized light in GAATR are similar to the absorption levels obtained by regular ATR in Figure 13.1. Thus, the enhancement phenomenon is limited to *p*-polarized light. However, even *s*-polarized light exhibits absorption levels that are several times stronger than the absorption levels that would be recorded if the silicon substrate was not present. The situation is depicted in Figure 13.2. Note that the difference between the two spectra is entirely due to the different optical materials behind the film. The difference for *p*-polarized light is even more dramatic than shown in Figure 13.1 if the angle of incidence is chosen to be the same. The gain due to the silicon substrate then climbs from 60 in Figure 13.1 to close to 100. This is an extraordinary gain in sensitivity. And, as we have seen, it is mostly expressed for *p*-polarized light, but there is a degree of enhancement that is also present for *s*-polarized light. Thus, we can conclude that there is a mechanism, somehow brought about by the presence of the silicon substrate, which boosts the electric field of the evanescent wave within the film by about an order of magnitude. However, although this is an unexpected phenomenon, nothing has to be added to the standard electromagnetic theory to account for it. We reproduced this phenomenon in Figures 13.1 and 13.2 by simply setting up the experimental parameters and by using the expression (Eq. 5.10) to calculate the resulting reflectance. This is yet another unexpected phenomenon that came out of a standard electromagnetic theory and expressions (Eq. 11.6).

What is the large enhancement of absorption due to? It looks like that, to have the enhancement, the film must be thin and the substrate must have a large refractive index.

13.3 ENHANCEMENT AND FILM THICKNESS

Let us first analyze the effect of film thickness on the enhancement. Obviously, as film thickness increases, the absorbance must increase since a larger fraction of the evanescent wave is then contained, hence absorbed, inside the film. For a thick film, the nature of the substrate must become irrelevant since the evanescent wave is localized near the interface. So for a thick film, the absorbance becomes independent of film thickness. Without the enhancement, the absorbance of the film would initially increase proportionally to the thickness than it would saturate and stay constant. However, the absorbance would never decrease with increasing film thickness. This situation is depicted in Figure 13.3 for $n_1 = 4$, $n_2 = 1.4 = 0.1i$, $n_3 = 1$, $\theta = 60°$ *p*-polarization.

As expected, the absorbance always increases with film thickness. If, for the same setup, only the substrate refractive index is changed from 1 (air) to 3.42 (silicon), the dependence changes rather dramatically, as shown in Figure 13.4. Note that already for the film thickness of 5000 Å (0.5 μm), absorbance for the thin film backed by air (Fig. 13.3) and the same film backed by silicon converges (Fig. 13.4) to the same value. So, although the wavelength of light used in the simulation is 10 μm, the bulk of the evanescent wave absorption occurs in the 0.5-μm-thick layer near the surface.

The change in behavior of absorbance with film thickness shown in Figure 13.4, as brought about by the change of substrate, is remarkable. With a silicon substrate, the absorbance rockets up almost five times above the value it would reach for an infinitely thick film. What that means is that, for a 500-Å-thick film, a bare 1/200th of the wavelength of light used (10 μm); the overall absorbance is five times higher than in the case of a thick film. The only way that can occur is if the electric field of the evanescent wave present inside the ultrathin film is much

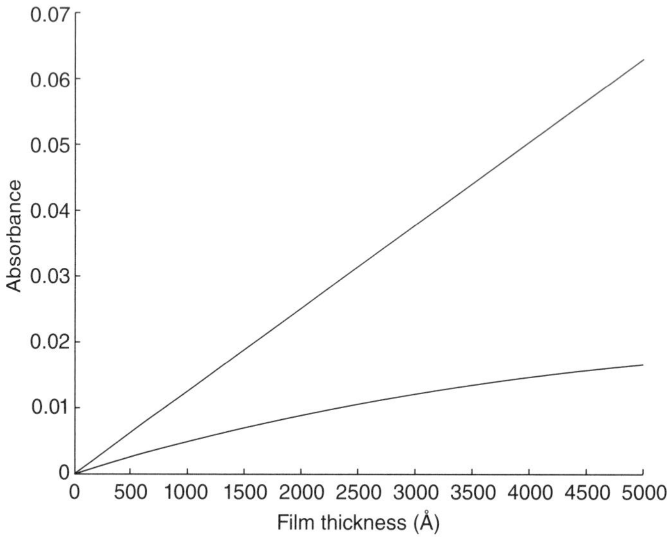

Figure 13.3 Absorbance of a thin film versus film thickness in ATR (lower curve) compared to the absorbance that the same film would exhibit in transmission.

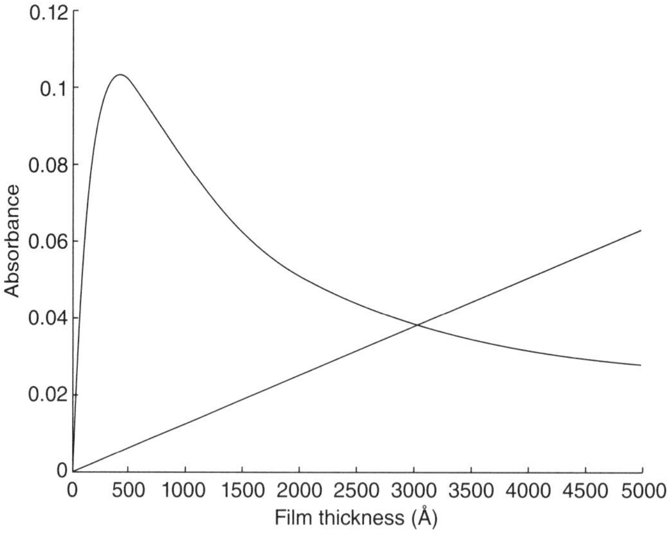

Figure 13.4 Absorbance of thin film versus film thickness (upper curve) for grazing angle ATR compared to absorbance of the same film in transmission.

more intense than the electric field of the evanescent wave for a thick film.

Since the absorbance is proportional to the product of the square of the electric field and the film thickness, we can estimate the field enhancement within the thin film compared to thick film as

$$d_f |E_f|^2 \approx 5 d_p |E_t|^2, \tag{13.1}$$

where E_f is the electric field of the evanescent wave inside the d_f thick film, d_p is the penetration depth for a thick film, and E_t is the electric field of the evanescent wave in the thick film. For our set of parameters $d_f = 0.05\ \mu\text{m}$ and $d_p = 0.5\ \mu\text{m}$, we get $|E_f|^2 = 50|E_t|^2$ or $E_f = 7E_t$. This is quite an enhancement. This enhancement grows even bigger for thinner films.

Why is the electric field of the evanescent wave inside an ultrathin film so much enhanced with respect to the electric field of the evanescent wave in a thick film?

13.4 ELECTRIC FIELDS IN A VERY THIN FILM ON A SI SUBSTRATE

To find the electric field inside the film, we take the same approach that led to the expression for the multilayer (Eq. 11.26). For a clearer analysis, let us assume that the angle of incidence is just below the critical. Since the reflection is subcritical, we have an internal wave, E_+, propagating toward interface 23 and a reflected wave E_- reflected from interface 23, returning to interface 12. The electric field inside the film, at depth z from the interface, is the sum of the electric fields of the two waves:

$$E_f(z) = E_+(z) + E_-(z). \tag{13.2}$$

Since the film thickness is so miniscule when compared with wavelength, the exponential term that describes the variation of the field with depth z is essentially the same everywhere inside the film. Thus, the dependence on z in Equation 13.2 could be ignored and dropped out. We can express E_- in terms of E_+ since E_- is just E_+ reflected at interface 23; that is,

$$E_- = r_{23} E_+. \tag{13.3}$$

At interface 12, the electric field of the incident wave E_0 is related to the internal fields as

$$E_+ = t_{12}E_0 + r_{21}E_-. \tag{13.4}$$

This allows us to express the internal fields in terms of the incoming field as

$$\begin{aligned} E_+ &= \frac{t_{12}}{1+r_{12}r_{23}}E_0 \\ E_- &= \frac{r_{23}t_{12}}{1+r_{12}r_{23}}E_0. \end{aligned} \tag{13.5}$$

Note that the two internal fields propagate at angles to each other so the electric fields of the two waves are not collinear. We first focus on the *p*-polarized incident light. For the component of the internal field perpendicular to the interface, we multiply the fields (Eq. 13.5) by

$$\sin\varphi = \frac{n_1}{n_2}\sin\theta$$

and add them together. Thus, the component of the internal electric field perpendicular to the interface, that is, the *z*-component, is

$$E_z = \frac{t_{12}}{1+r_{12}r_{23}}(1+r_{23})\frac{n_1}{n_2}\sin\theta E_0. \tag{13.6}$$

For the component of the internal field parallel to the interface, that is, the *x*-component, we multiply the fields (Eq. 13.5) by

$$\cos\varphi = \frac{\sqrt{n_2^2 - n_1^2\sin^2\theta}}{n_2}.$$

However, since in our choice of geometry these two components are pointing opposite to each other, they have to be subtracted. Thus, the *x*-component of the internal electric field parallel to the interface is

$$E_x = \frac{t_{12}}{1+r_{12}r_{23}}(1-r_{23})\frac{\sqrt{n_2^2 - n_1^2\sin^2\theta}}{n_2}E_0. \tag{13.7}$$

13.5 SOURCE OF ENHANCEMENT

The amplitude coefficients in Equations 13.6 and 13.8 are both for the p-polarized incident light. Using the values $n_1 = 4$, $n_2 = 1.4$, $n_3 = 3.42$, $\theta = 60°$, we find $r_{23} = 0.9434$. This makes the parallel field component (Eq. 13.7) very small. If we use the above values for n_1, n_2, n_3, and θ, we find

$$\left|\frac{E_z}{E_0}\right| = 13.2308.$$

This is a remarkable enhancement of the perpendicular component of the electric field inside the film.

The absorbance is proportional to the square of the magnitude of the field. This is consistent with the huge enhancement that we saw in Figure 13.4.

For the x-component, we find

$$\left|\frac{E_x}{E_0}\right| = 0.3527,$$

which, after squaring, becomes negligible.

The analysis for the s-polarized light is simpler since the two fields are collinear, parallel to the interface, and along the y-axis. Again, using our values for n_1, n_2, n_3, and θ, we find

$$\left|\frac{E_y}{E_0}\right| = 1.9282.$$

The electric field of the evanescent wave for s-polarized incident light is certainly not suppressed. It only seems so in comparison to the strength of p-polarized light, which is 5–10 times stronger.

We have seen that the component of the electric field of the evanescent wave perpendicular to the interface is remarkably enhanced, while the two parallel components are not. We have also seen that this enhancement is high for the extremely thin films, and also, that the enhancement requires a high refractive index substrate. To understand this phenomenon a bit more, let us examine how the value of $|E_z/E_0|$ in this case compares to the same expression for the case of a thick film. For a thick film, there is no E_- wave since the evanescent wave never reaches the 23 interface. We can simulate this by setting $r_{23} = 0$.

The expression in Equation 13.6 then reduces to the expression in Equation 6.4b that we encountered earlier. We already saw in Equation 6.4 and in Figure 6.3 that, for a single interface, E_z is maximized at the critical angle and goes to zero as the angle of incidence increases, and that E_x is zero at the critical angle, grows to a maximum at some angle between the critical and 90°, and then also goes to zero at the critical angle.

For a single interface, the normal component of the electric field at the critical angle is

$$\left|\frac{E_z}{E_0}\right| = 2\frac{n_1^2}{n_2^2}\sin\theta = 2\frac{n_1}{n_2}, \tag{13.8}$$

which, for the above choice of refractive indices, numerically evaluates to 5.71. Thus, the enhancement in the field strength is due to the high ratio of refractive indices n_1/n_2. This is then quenched by a relatively small sine of the critical angle. Note that $\sin\theta$ is necessarily small for a high refractive index ATR material. While not a small enhancement, the result (Eq. 13.8) shows the limitation to enhancement achievable with a single interface. A large factor, $2(n_1^2/n_2^2)$, is tempered by a small sine of the critical angle $\sin\theta = n_2/n_1$. If the angle of incidence is increased, the sine grows, but the transmission amplitude coefficient plunges toward zero. So, what happens with an ultrathin film on a silicon substrate is that the critical angle is increased, replacing n_2 by n_3 in the numerator of the sine. At the same time, the E_- contribution is added, virtually doubling the magnitude of the field.

What was just said did not involve film thickness. We know (Fig. 13.4) that film thickness plays a crucial role in the enhancement. Let us then see how the key ratio $|E_z/E_0|$ changes with film thickness. We have to take into account Equations 11.16 and 11.18 to find the expression that is the counterpart of Equation 13.6:

$$E_z(0) = \frac{t_{12}e^{4\pi ikd\sqrt{n_2^2-n_1^2\sin^2\theta}}}{1+r_{12}r_{23}e^{4\pi ikd\sqrt{n_2^2-n_1^2\sin^2\theta}}}(1+r_{23})\frac{n_1}{n_2}\sin\theta E_0. \tag{13.9}$$

Note that Equation 13.9 reduces to Equation 13.6 in the limit when $d = 0$. We can now calculate the magnitude of the field component E_z at the interface as a function of film thickness. The result is shown in Figure 13.5. We can see the steep drop-off in field intensity as a function of the film thickness. The end of the thickness scale is a mere 1/20th of the wavelength.

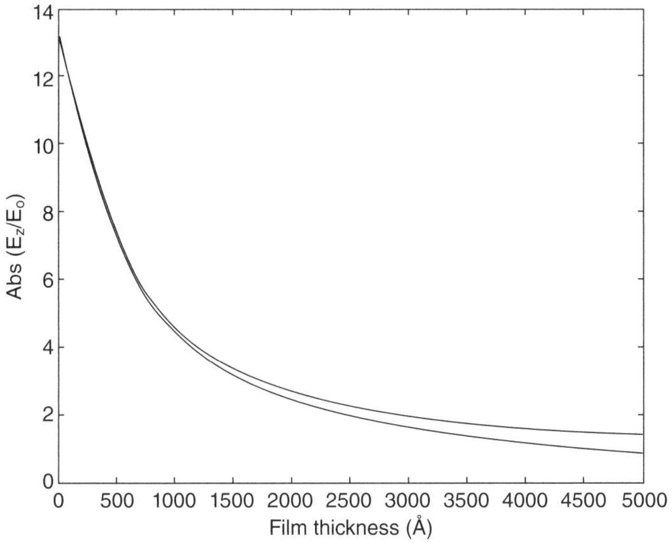

Figure 13.5 Magnitude of the field component E_z at the interface as a function of film thickness. For comparison, the magnitude of E_z at the second interface (lower curve) is also shown.

The field strength does not change much within the film as the lower line on the graph in Figure 13.4, which represents the field strength on the opposite interface, indicates.

Note that the only difference between the expressions in Equations 13.6 and 13.9 is the presence of the exponential term. The values of the exponent for $d = 0$ cm and $d = 5 \cdot 10^{-5}$ cm are 1 and 2.5, respectively, so the exponential term ranges from 1 to 0.082, quite a considerable change.

13.6 GAATR SPECTROSCOPY

A note on collecting a GAATR spectrum is in order. If the film material is nonabsorbing, the reflectance of such a system is total, regardless of the film thickness. Therefore, when recording a spectrum in GAATR, one collects a background spectrum using a germanium ATR element without anything in contact with it. We know that if we had a nonabsorbing film, with a refractive index identical to the sample, which was deposited on a silicon substrate, the resulting reflectance would still be total as it is with the germanium ATR element alone. This is guaranteed by the laws of total internal reflection. So, we can then proceed and collect the sample spectrum by placing the sample over the ATR crystal. Any absorbance thus recorded is entirely due to the thin film.

It is interesting to compare how the absorbance produced by the GAATR compares with the absorbance produced by transmission spectroscopy for the same film. We can accomplish that by calculating the ratio of the two spectra from Figure 13.4. The result is shown in Figure 13.6.

The most striking insight that follows from Figure 13.6 is that the boost over transmission for GAATR is the highest for zero film thickness and drops off precipitously with increasing thickness. The maximum boost is just above 50X. This is a significant enhancement to sensitivity for thin films, but we see that the results of a GAATR experiment must be interpreted cautiously, especially in the case when the film thickness is not known.

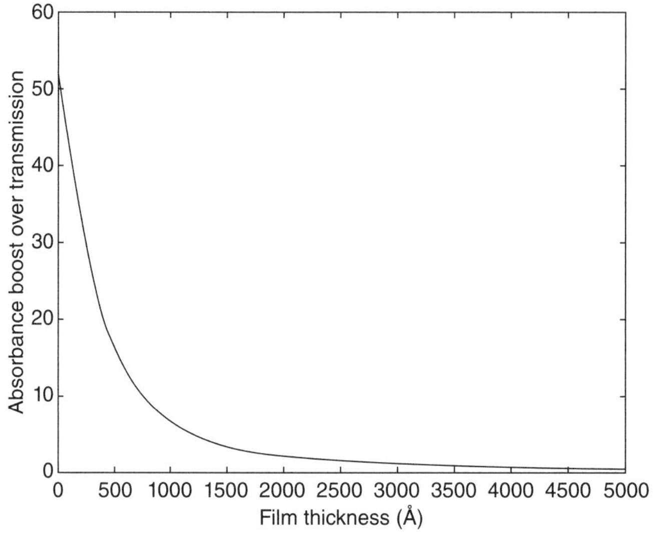

Figure 13.6 Absorbance boost over transmission exhibited by grazing angle ATR.

14 Super Grazing Angle Reflection Spectroscopy (SuGARS)

14.1 REFLECTANCE OF THIN FILMS ON METAL SUBSTRATES

Naturally, the subject of thin films on silicon substrates is of great interest to the semiconductor industry. Another substrate that yields a similarly strong enhancement for ultrathin films deposited on it is metal. A number of different metals could be used, but gold is a popular substrate due to its high reflectivity and superior chemical inertness.

So let us now turn to thin films on metal substrates. First, as we have already seen (Fig. 12.4), the interface between an ATR element and metal is actually not an internally reflecting interface. Thus, the reflection is not total. Second, in regular grazing angle specular reflection, light refracts at the air–film interface. Thus, although light may have been incident on the film surface at a very high angle of incidence, say, 80°, it refracts into the film and, for a typical film material with a refractive index around 1.40, it propagates through the film at about 45°. It is impossible for light to propagate within the film at a much higher angle than 45° regardless of how high the angle of incidence is. Therefore, the path length through the film is limited to about three times the film thickness. This limits the sensitivity of grazing angle specular reflection spectroscopy.

However, if light is incident from a material with a refractive index larger than that of the film, this limitation is no longer present. For instance, if light is incident from a material with the same refractive index as the film, light does not refract at the interface but propagates through the film at the angle of incidence. Thus, the total path length through the film becomes $2d/\cos\theta$, where θ is the angle of incidence, which can be anything up to 90°. For $\theta \rightarrow 90°$, $\cos\theta \rightarrow 0$; thus, the path length through the film diverges to infinity, bringing the sensitivity with

Internal Reflection and ATR Spectroscopy, First Edition. Milan Milosevic.

it. At the grazing angle, light propagates down the film essentially parallel to the metal surface. If the incident medium has a refractive index larger than the film, the propagation down the film parallel to the interfaces is reached as the angle of incidence onto the film becomes the critical angle for the medium–film interface. For angles of incidence above this critical angle, light exists in the film only as an evanescent wave. This type of reflection could thus be termed super grazing reflection and the spectroscopy associated with it super grazing angle reflection spectroscopy (SuGARS).

14.2 PROBLEM OF REFERENCE

What is the proper configuration for collecting the background spectrum in the case of a metal substrate? A clean ATR element? An ATR element covered with a clean metal substrate? Or something else? Let us proceed by first observing the dependence of absorbance on film thickness for the same film as with the silicon substrate, but this time backed with the metal substrate instead. The calculated film + substrate reflectance is ratioed to the reflectance calculated with the substrate alone. The results for the two different angles of incidence are shown in Figure 14.1. Transmission results are shown for comparison.

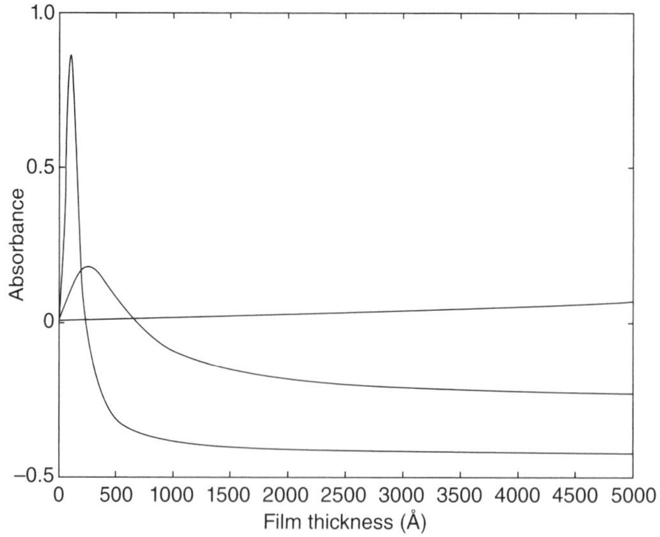

Figure 14.1 Absorbance of a thin film on a metal substrate versus film thickness For reference, the transmission of the same film for normal incidence is shown (linearly rising curve). The sharply peaked curve is for the angle of incidence of 80° and the other curve for 60°.

Two observations are in order here. First, note the really high absorbance for ultrathin films at an 80° angle of incidence. Second, note the unexpected appearance of the negative absorbance for thicker films. While the high absorbance for thin films is good news in the quest for high sensitivity, the presence of negative absorbance is anticlimactic. Negative absorbance does not make sense, and we must understand its origin. First, negative absorbance, in this context, simply means that the reflectance that was transformed into absorbance was greater than one. Again, this does not make sense. However, what it really means is that the reflectance of the metal substrate was improved by depositing a thin film on it even though the film is somewhat absorptive.

Now we see that the negative absorbance is not that surprising. It is actually standard practice to deposit thin films over metal surfaces in order to improve their reflectivity. If, for the background spectrum, we use the reflectance spectrum of a thin nonabsorbing film on the metal substrate that has the same thickness and refractive index as the absorbing film, we would avoid negative absorbance as shown in Figure 14.2. The graph from Figure 14.2 shows the same high sensitivity, but the negative absorbance is gone.

The result is not too comforting, however. What we did to remove negative absorbance is easy enough to do in a numerical simulation but extremely hard to do in an actual experiment. The result shown in Figure 14.2 at least confirms that we fully understand the source of

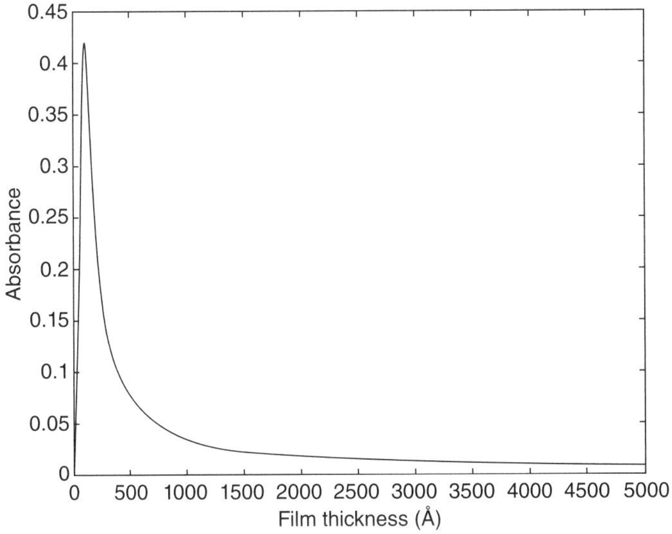

Figure 14.2 Absorbance versus film thickness for super grazing angle reflection spectroscopy with a nonabsorbing film with the same thickness and refractive index used for the background spectrum.

negative absorbance from Figure 14.1. Thus, the SuGARS technique is very useful for ultrathin films, where it provides remarkable sensitivity, but the usability of this technique for thicker films, and in this context thicker encompasses anything thicker than about 50 Å, is marginal, or at best only qualitative.

14.3 SENSITIVITY ENHANCEMENT

As we did for grazing angle ATR, we can quantify the boost in sensitivity that SuGARS gives over normal incidence transmission spectroscopy. We can calculate the ratio of the absorbance produced by SuGARS and the absorbance produced by normal incidence transmission for the same film thickness.

The result is shown in Figure 14.3. As with grazing angle ATR, the boost is also thickness dependent, but this time, the maximum does not occur for zero film thickness. It is shifted to a higher film thickness. Where exactly the maximum boost occurs is dependent on the angle of incidence and the properties of the two materials. For an angle of incidence of 80°, the maximum is reached at about 100 Å where the boost reaches an astonishing value of 330. While the details change with specific material parameters, the conclusion regarding the extraordinary sensitivity of SuGARS remains generally valid. Note that the high sensitivity can be extended to thicker films by somewhat decreasing the angle of incidence, but at the expense of sensitivity for the ultrathin films.

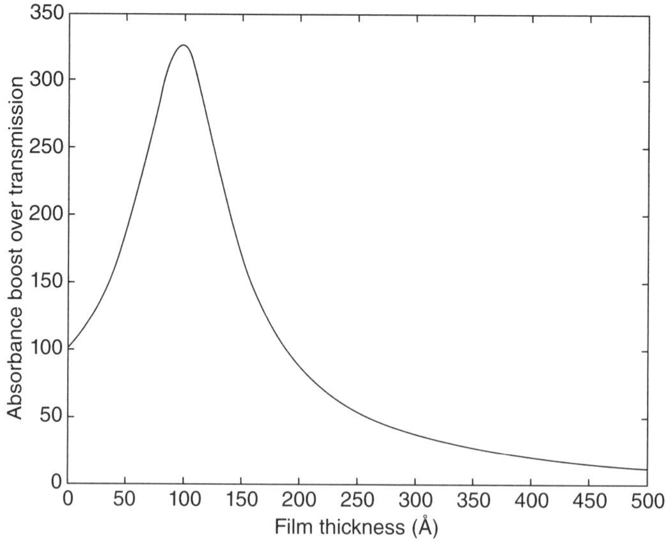

Figure 14.3 Absorbance boost over transmission exhibited by SuGARS.

15 ATR Experiment

15.1 MULTIPLE REFLECTION ATTENUATED TOTAL REFLECTION (ATR)

We have already seen how a spectrometer made for transmission measurements can be used to measure the reflectance of a sample. The setup requires the use of a reference reflector so that the unknown reflectance of a sample can be expressed as a fraction of the reflectance of a known reference. The ATR technique brings a twist to this procedure. The internally reflecting surface of the ATR element is totally reflective in the absence of a sample. This is the ideal reference, but this benefit comes with a side effect, namely, light has to enter and exit the ATR element, and the entrance and exit facets of the element introduce their own reflectance into the picture. And since the refractive index of ATR materials is necessarily high, the reflectance of these facets is all but negligible. As a consequence, the light exiting the ATR element consists of multiple components, each imprinted with a different number of reflections within the element. Therefore, while the reflectance of a sample is indeed imprinted onto the exiting light, it is not imprinted in a simple way. We need to analyze the details of light going through the ATR element in order to understand how to untangle the actual sample reflectance from the result obtained by the reflectance measurement procedure.

The geometry of light traversing a typical multireflection ATR element is shown in Figure 15.1. The angle of reflection shown is 45°, but it could be any supercritical angle. The multiple reflection ATR element shown has the so-called trapezoid shape, which is very popular since the longer face of the element can be oriented to be horizontal and above the optics needed to bring light in and out of the element. The horizontal configuration makes sample introduction and cleanup very convenient.

Internal Reflection and ATR Spectroscopy, First Edition. Milan Milosevic.

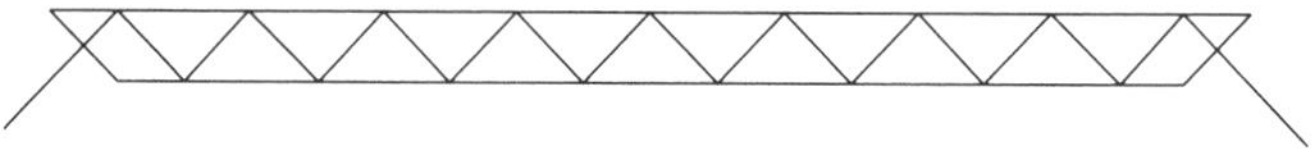

Figure 15.1 Geometry of multireflection ATR element.

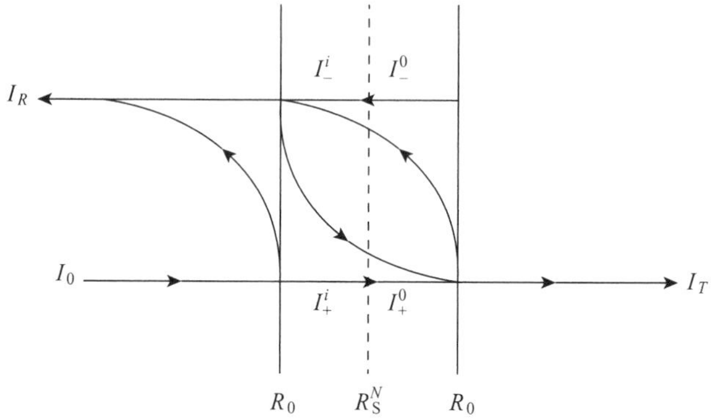

Figure 15.2 Simplified geometry of multiple ATR.

Assume that light is incident from left onto the entrance facet. The entrance facet is cut to provide normal incidence at the facet so that light transmitted through the facet retains the direction of incident light. Transmitted light zigzags through the element experiencing reflections on the top and bottom surfaces of the ATR element and arrives to the exit facet wherein it partially transmits out of the element and partially reflects back. This back-reflected component retraces the original path through the element but in reverse. As it comes to the exit surface, a component transmits through and a component reflects back and travels one more time through the element experiencing additional reflections, and so on.

Thus, light exiting the ATR element contains components that experienced a different number of reflections. Note that the number of reflections on the top surface is always one more than the number of reflections on the bottom surface. We assume that the sample is applied only to the top surface and that there are N reflections on the top surface.

We can simplify the above geometry into the one shown in Figure 15.2. We denote the reflectance of the entrance (and exit) facet as R_0 and the internal reflectance of the sample–ATR interface as R_s. The dashed line in Figure 15.2 represents the N internal reflections on the sample–ATR element interface. Inside the ATR element, there are two streams of light, I_+ traveling toward the exit aperture and I_- traveling in the opposite direction. As these components traverse the element (in either direction), they pick up the factor R_s^N. So they have different intensities at the exit facet than they have at the entrance facet.

From Figure 15.2 it is clear that

$$\begin{aligned} I_+^0 &= R_s^N I_+^i \\ I_-^i &= R_s^N I_-^0. \end{aligned} \tag{15.1}$$

It follows that

$$\begin{aligned} I_R &= R_0 I_0 + (1 - R_0) I_-^i \\ I_T &= (1 - R_0) I_+^0 \end{aligned} \tag{15.2}$$

and that

$$\begin{aligned} I_+^i &= (1 - R_0) I_0 + R_0 I_-^i \\ I_-^0 &= R_0 I_+^0. \end{aligned} \tag{15.3}$$

By combining Equations 15.1–15.3, we find for the transmitted light

$$T = \frac{I_0}{I_T} = \frac{(1 - R_0)^2 R_s^N}{1 - R_0^2 R_s^{2N}}. \tag{15.4}$$

Note that we derived the expression (Eq. 15.4) using an approach very similar to deriving the reflectance and transmittance of multilayers (Eq. 11.28). This is not surprising since we are essentially looking for the same quantities here as we were looking for in the multilayer case. However, here, we did not use wave amplitudes; we used the intensities directly. Why were we allowed not to worry about the wave nature of light? The answer is that, here, the separations between the various interfaces are enormous when compared to the wavelength of light. So, the propagation terms (Eq. 11.16) that we used to connect amplitudes at the neighboring interfaces become so highly oscillatory that even the tiniest angular deviation from the nominal direction makes the oscillatory terms have any value between 1 and –1. Thus, in taking the square of amplitude to obtain the intensity of the transmitted light, only the product of each component with itself survives. All the products of different components oscillate wildly between 1 and –1 and hence average to zero. The wave nature of light for thick films in effect disappears for all but extremely collimated beams such as laser beams. The measured quantity T contains the desired quantity R_s, but it is part of a complicated expression (Eq. 15.4). Furthermore, the quantity R_0 is not negligible since an ATR material necessarily must have a high refractive index in order to be suitable for ATR spectroscopy. Normal incidence reflectance is from Equation 4.17a:

$$R_0 = \frac{(n-1)^2}{(n+1)^2}. \tag{15.5}$$

For example, germanium has a refractive index of 4. According to Equation 15.5, for a germanium ATR element, we have $R_0 = 0.36$. This is certainly not negligible. Note further that it follows from Equation 15.4 that, in the absence of the sample (i.e., $R_s = 1$), the transmitted intensity is

$$T = \frac{(1-R_0)^2}{1-R_0^2}. \tag{15.6}$$

Thus, the measured quantity is not Equation 15.4 itself but Equation 15.4 divided by Equation 15.6; that is,

$$R = \frac{(1-R_0^2)R_s^N}{1-R_0^2 R_s^{2N}}. \tag{15.7}$$

We see from Equation 15.7 that, if the facet reflectance vanishes (for instance, with the help of an antireflection coating), the expression reduces to R_s^N. On the other hand, R_0 is known through Equation 15.5. Therefore, it is not difficult to express the value that we want (i.e., R_s^N) in terms of the measured quantity R and the facet reflectance R_0 as

$$R_s^N = \frac{\sqrt{(1-R_0^2)^2 + 4R^2R_0^2}}{2RR_0^2}. \tag{15.8}$$

This is an interesting result. It shows that the ATR technique actually enables the measurement of the absolute internal reflectance of the sample. Unlike with regular reflectance, where sample reflectance is expressed in terms of the reflectance of a reference, the ATR technique takes advantage of the total reflectance of the interface in the absence of a sample to provide the true (absolute) reflectance of the sample. However, the need for the reference reflector is traded for the presence of the facet reflectance. The facet reflectance is easy to find since the refractive indices of all ATR materials have been measured to a high precision for the entire range of transparency of each material. To measure the external reflectance of a sample, the sample surface must be made flat and polished. This may not be easy, or even possible, with many samples. However, pressing a sample into contact with an already flat and optically polished surface of an ATR element essentially imprints the flatness and optical polish of the ATR element onto the sample. Therefore, measuring the absolute reflectance of liquids, pastes, and soft solids could become routine. Note that the result (Eq. 15.8) remains valid for any number of reflections, and particularly for one reflection since one reflection ATR is of significant practical importance.

15.2 FACET REFLECTIONS

We can examine the importance of facet reflections by way of an example. Figure 15.3 compares a true ATR spectrum of a typical sample with the spectrum that would be measured using a single reflection germanium ($n = 4$) ATR element. The angle of incidence is 45° and the

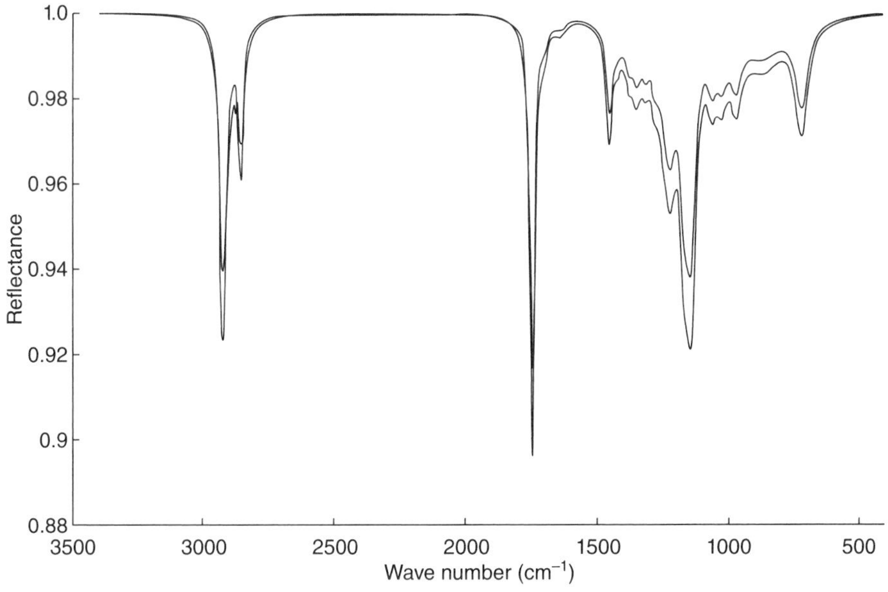

Figure 15.3 The ATR spectrum of a sample as measured on a germanium ATR element compared to the true spectrum of the sample (upper curve).

polarization of the incident beam is p (parallel to the plane of incidence). We see that the effect is not small and clearly not negligible.

In almost all applications of ATR spectroscopy, peak strengths themselves are not the goal of the measurement. Rather, the peak strength is used as a measure of the concentration of the particular substance responsible for the peak. A set of standards with known concentrations of the substance of interest is prepared; spectra are measured; and the peak strength is correlated to the concentration. A curve (preferably linear) is fitted through the points and the functional relationship between the concentration and the peak strength (peak height, area under the peak, etc.) is extracted. If now an unknown concentration of this substance in the same matrix needs to be determined, the measured peak strength is used to calculate the unknown concentration.

Therefore, the actual details of a spectroscopic measurement, such as the angle of incidence, the number of reflections, the refractive indices of the sample and the ATR material, and the polarization of incident light, are not needed at all to obtain the unknown concentration. This is an important point and we will come back to further discuss it later on.

The actual spectrum of the sample (Figure 15.3) is "amplified" through the effect of facet reflections. This is so because the reflected components have passed through the crystal multiple times, each time experiencing additional reflections, so the effective number of reflections inside the ATR element is more than the nominal number suggested by the geometrical considerations.

This closely parallels a similar effect that takes place in transmission spectroscopy. A sample, say, a solid film, reflects the transmitting light at both its front and back surface. The back-reflected light from the second surface travels back through the sample, partially reflects from the front surface, passes through the sample, reaches the back surface where it again partially exits and partially reflects back, continuing the multiple reflections. Although most samples have low refractive indices and a very small fraction of incident light reflects from each surface, the effect is far from negligible as we have learned from studying the phenomenon of interference fringes.

15.3 BEAM SPREAD AND THE ANGLE OF INCIDENCE

A typical beam of a Fourier transform infrared spectrometer is not collimated within the spectrometer's sample compartment. Rather, it is focused into the sampling spot inside the sample compartment. The

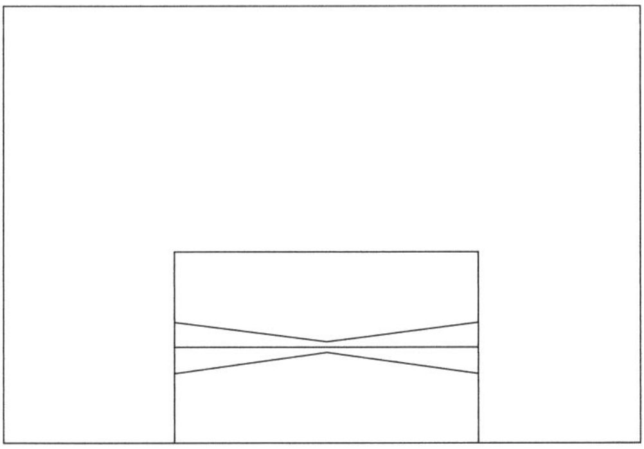

Figure 15.4 Top view of a typical spectrometer. An outline of a typical optical beam in the sample compartment is shown. The central ray of the beam is indicated as well as the extreme rays of the beam. The waist of the beam is the focal point where samples are normally introduced for transmission measurements. The rays of the beam fill more or less uniformly the entire conical volume centered on, and axially symmetric around, the central ray.

spot size is about 6 mm in diameter and the rays in the beam have a beam spread of about ±7°. This is illustrated in Figure 15.4.

Other types of optical spectrometers that use gratings to disperse light have similar beam spreads and spot sizes, but the beam in the focal point is slit shaped rather than round. We can clearly see from the beam geometry that if a sample is placed in the beam so that the central ray is normal to the sample surface, there are infinitely many more rays that are at a small angle to the normal than there are rays that are exactly normal to the sample. Therefore, the average angle of incidence cannot be zero but must be somewhat larger than zero. The rays in the beam that are not normal (and there is really only an infinitesimal fraction of the rays in the beam that are truly normal) do not have their individual path lengths through the sample equal to the sample thickness, but, because they are slightly slanted, the path lengths are somewhat larger than the film thickness.

There is an analogue of the same effect in ATR. Figure 15.1 shows light propagating through a multireflection crystal. Strictly speaking, the line representing the path of light is true for the central ray of the beam but not for any other ray in the beam. Each individual ray in the beam enters the ATR element at its own angle, thus generating a distribution of angles of incidence. With a single reflection ATR element, this distribution of angles of incidence does not have any further effect. However, for a multiple reflection ATR, the change in the angle of incidence may lead to a change in the number of reflections. Thus, for multiple reflection ATR elements, in addition to the distribution of angles of incidence, we are also faced with a distribution of a number of reflections that various rays in the beam experience on their way through the ATR element.

15.4 EFFECT OF FACET SHAPE

Let us analyze the several ways in which light can be introduced into an ATR element. First, let us observe the effect of the entrance facet on the beam spread and the effect on the size and position of the beam's focal point. Figure 15.5 shows the two most common shapes of entrance facets. The beam is incident from the left side. The flat facet (left) shows the beam converging to the focal point. The rays refract on the surface with the effect that the focal point is pushed further from the interface. Simple geometrical considerations show that the beam is pushed forward by $(n - 1)d$, where d is the distance of the beam's original focal point from the interface and n is the refractive index of the ATR material. The beam spread is consequently reduced. However, the size of the beam in the focal point is not changed.

This is contrasted with the case where the entrance facet is spherical with the center of curvature coincident with the original beam focus.

Again the beam refracts at the surface. However, the consequences of this are somewhat peculiar. First, we see that the choice that the center of curvature is coincident with the beam's original focal point has for a consequence that the refraction does not alter the position of

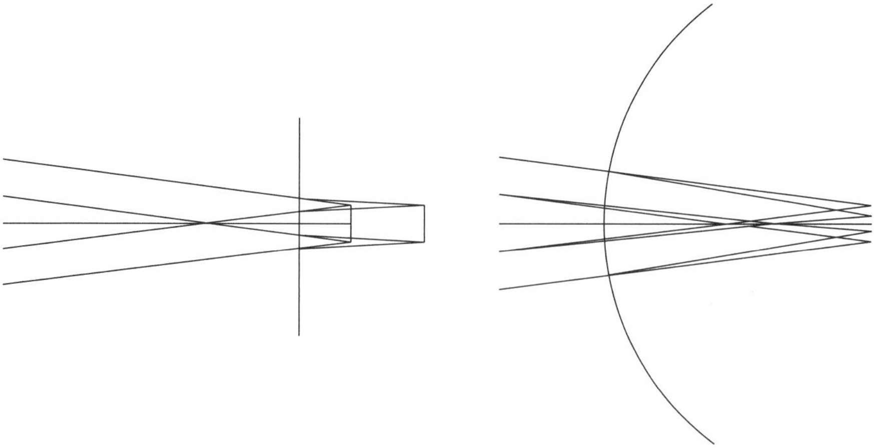

Figure 15.5 Effect of the curvature of the entrance facet on the beam inside the ATR element.

the focal point. Another peculiar effect is that, although the beam position did not change, the beam size in the focal point is reduced. Simple geometrical considerations show that the beam size is n times reduced where n is the refractive index of the ATR material. Because ATR materials have very high refractive indices, the reduction in beam size is significant. For germanium with $n = 4$, the spot size is reduced an astounding four times. This beam spot size reduction is often erroneously described in the literature on ATR microspectroscopy as akin to oil immersion often employed by high-power microscope objectives to further increase the magnification of the objective. Unlike the above described effect, immersion imaging does not rely on the object being positioned in the center of a spherical surface, and the position of the object changes with oil immersion.

There is some beam spread increase due to this spot size reduction, but this is generally negligible for situations where beam size is much smaller than the radius of curvature of the spherical surface. The ATR elements employing the beam geometries from Figure 15.5 are shown in Figure 15.6.

The spot size reduction effect of a spherical surface discussed above is obviously of great interest in those ATR applications where the sample itself is very small or where a small sample is embedded in a larger matrix. The beam size reduction brings all the intensity of the incident beam into a smaller spot essentially without a corresponding increase in the beam spread.

The image size reduction, although caused by the refraction on a spherical surface, which is controlled by the refractive index of the ATR element, is still almost perfectly achromatic since the position of the image is in the center of the sphere regardless of the value of the refractive index. The qualifier "almost" above applies to the image size—not the position. The size is directly controlled by the index of refraction and it changes as the refractive index changes for different wavelengths.

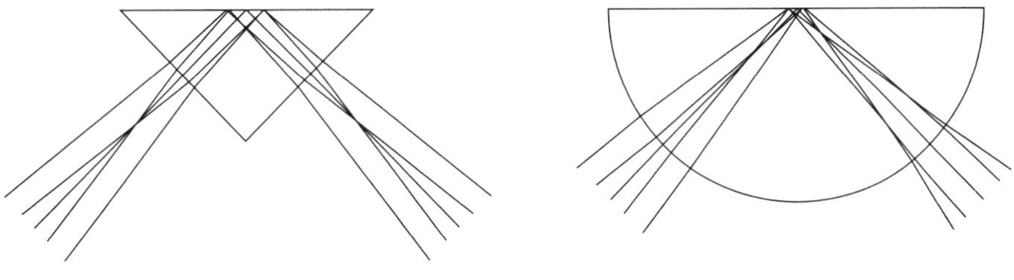

Figure 15.6 ATR elements employing the beam geometries from Figure 15.5. The beam paths for different rays shown were drawn following the laws of refraction and reflection.

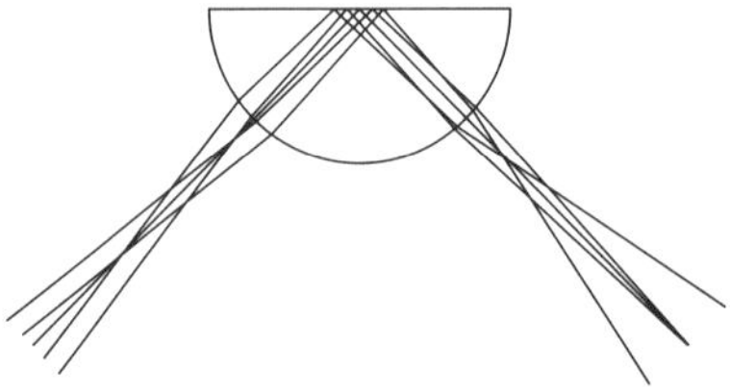

Figure 15.7 ATR geometry that minimizes the beam spead inside the ATR element.

There is another important variation of using spherical facets to affect the beam spread inside ATR elements. This is illustrated in Figure 15.7. The focal point of the beam is positioned outside the ATR element at a distance d from the spherical surface. The refraction at the spherical surface then creates a collimated beam inside the ATR element. As the collimated beam exits the ATR element, the exit spherical surface refocuses the beam back at the same distance d from the spherical surface. Simple geometrical considerations show that this distance is given by $d = R/(n - 1)$, where n is the refractive index on the ATR material.

The situation shown in Figure 15.7 depicts the use of this type of an ATR element with a typical spectrometer beam. A better way to utilize this type of geometry would be to use a beam that is faster and with a smaller spot size since this type of imaging works best when the size of the beam in the focal point is much smaller than the radius of curvature of the ATR element. However, even with the nonideal arrangement of Figure 15.7, it is clear that the configuration shown works well to minimize the beam spread inside the ATR element. The above configuration provides a good way to have a well-defined angle of incidence in ATR measurements. The trade-off is that this method requires a larger sample size in contact with the ATR element since here the beam is not focused at the sampling surface. Thus, one can choose between the configuration that provides a small sample spot and a large spread in the angle of incidence (right side in Fig. 15.6), or the configuration that provides a small spread in the angle of incidence and a large sample spot (Fig. 15.7).

15.5 BEAM SPREAD AND THE NUMBER OF REFLECTIONS IN MULTIPLE REFLECTION ATR

Let us now turn to multiple reflection ATR elements. As we have already seen, the situation with multiple reflection ATR is more

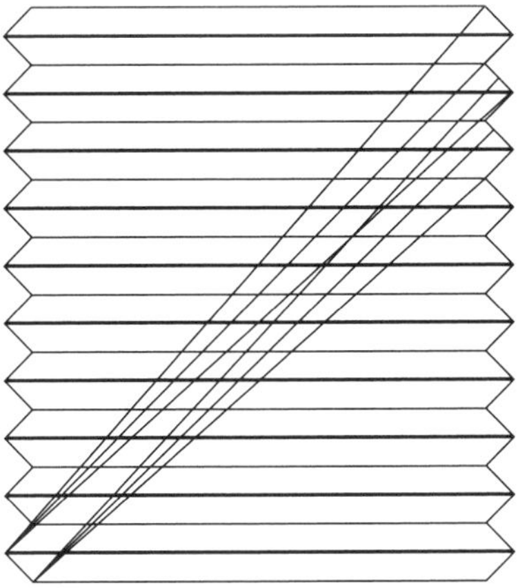

Figure 15.8 Geometry of light propagation through multiple reflection ATR.

complicated. Figure 15.8 illustrates the spectrometer's beam entering the multiple reflection ATR element of Figure 15.1 on the bottom left side. Several rays are shown indicating the beam spread inside the element. Upon entering the element, the beam spread is reduced through Snell's refraction. Figure 15.8 shows the refracted beam with the reduced beam spread. The beam first reflects from the top then the bottom surface, and so on, as it zigzags through the element toward the exit facet. With all the reflections shown, the actual picture quickly becomes intractable. Each reflection from a plane surface viewed as drawn on a piece of paper could be seen as a folding of the paper at the plane of the reflection. Thus, we can imagine the paper unfolded first on the top surface, then on the bottom, and so on. The result of this unfolding is shown in Figure 15.8.

If we apply the sample to the top surface only, then each time a ray crosses a heavy horizontal line (an unfolding of the top surface), the ray experiences one reflection. There are 10 heavy horizontal lines in Figure 15.8. We see that some rays make it all the way through the tenth reflection line and exit through the exit facet above the tenth line. No rays reach the eleventh reflection line. The rays that crossed the tenth reflection line have experienced 10 reflections. A portion of the rays that made it through the ninth reflection line, but not through the tenth reflection line, has experienced nine reflections and exits through the facet above the ninth reflection line. Similarly for the rays that exit after passing through the eighth reflection line. Finally, there is a small fraction of rays that exit after seven reflections.

Thus, clearly, not all rays experience the same number of reflections. We can label the rays according to the number of reflections they experience. Let us label the fraction of rays that experience N reflections as a_N. Then, the intensity of light that exits the ATR element can be expressed as

$$I_R = \left(\sum_N a_N R^N\right) I_0. \qquad (15.9)$$

Therefore, the "transmittance" T of the ATR crystal is

$$T = \frac{I_R}{I_0} = \sum_N a_N R^N. \qquad (15.10)$$

If all the rays underwent the same number of reflections, the equivalent of Equation 15.10 would be

$$T = R^N. \qquad (15.11)$$

The absorbance transform of Equation 15.10 yields

$$A = N \cdot \log(R); \qquad (15.12)$$

that is, the absorbance seen in the multiple reflection ATR would be the absorbance seen in one reflection ATR multiplied by the number of reflections. This is no longer true when different rays in the beam undergo different numbers of reflections. We could still attempt to define the effective number of reflections in analogy with Equation 15.12 as

$$N_{\text{eff}} = \frac{\log(T)}{\log(R)}. \qquad (15.13)$$

15.6 EFFECT OF BEAM ALIGNMENT ON MULTIPLE REFLECTION ATR

Let us illustrate the effective number of reflections by an arbitrary but realistic example. For the case from Figure 15.8, we could somewhat arbitrarily, but not unrealistically, write

$$T = 0.15R^{10} + 0.35R^9 + 0.45R^8 + 0.05R^7. \qquad (15.14)$$

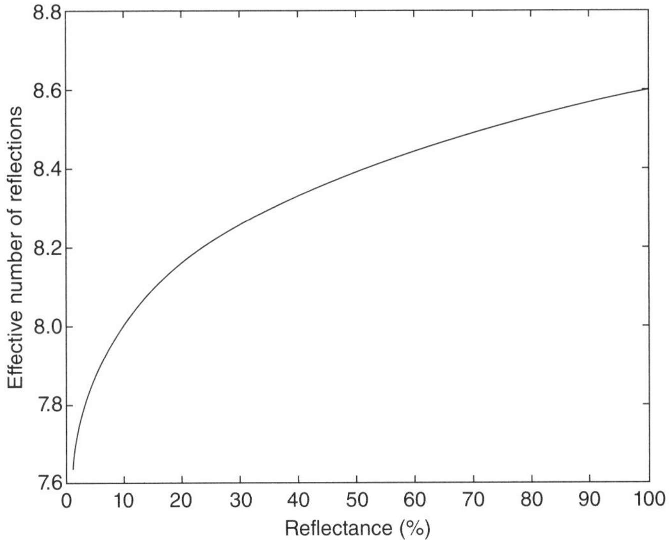

Figure 15.9 Effective number of reflections versus reflectance for multireflection ATR.

The result is shown in Figure 15.9. For a high sample reflectance, the effective number of reflections is around 8.6, but as the sample reflectance decreases, the effective number of reflections drops to 7.6. Since the reflectance drops with the increasing absorbance of the sample, the change in the effective number of reflections decreases with an increase in the absorbance, hence introducing a nonlinear relationship between multiple and single reflection ATR.

This phenomenon superimposes itself on top of the effect of facet reflections that we considered before. All these phenomena clearly complicate the extraction of the actual reflectance of the sample–ATR interface. But, as we have discussed earlier, this is rarely a concern in real-life applications of multiple reflection ATR spectroscopy. The purpose of most ATR measurements is to determine the concentration of an unknown component in a mixture. Could this nonlinearity have an effect despite the calibration procedure mentioned earlier?

For instance, imagine that you are asked, by a somewhat mischievous colleague, to measure the concentration of alcohol in a number of samples of wine. You would first prepare a set of standards where you would carefully add a known amount of pure ethyl alcohol to known amounts of water. In this way, the concentrations of alcohol in water in your standards would be known precisely. Then you would take the ATR spectra of your standards using your multiple reflection ATR unit. You would observe the absorption peaks associated with alcohol and you would select a peak and plot its intensity versus concentration. Even if the resulting curve is not linear, you would then be in a position to measure the alcohol content of the wines. Obviously, you would acquire the ATR spectrum of an unknown sample using the same apparatus and conditions that you used to measure the spectra of your standards. You would observe the peak height of the alcohol band, consult your calibration curve, and read out the concentration of alcohol.

Therefore, what you would be able to declare is that if you prepare yet another standard with the concentration that you just read from your calibration graph as the concentration of alcohol in your unknown sample, the alcohol peak intensity would closely match that of the unknown wine sample. Since wine is essentially just a mix of alcohol and water plus some minor components, you would also expect the water peaks of your new standard to match that of wine.

Some additional minor absorption bands in the spectrum of wine you would ascribe to those minor components in wine that make wine wine and not just a mix of water and alcohol. If everything proceeded as just described, you would be satisfied with your results and confident in your experimental method. You would measure the alcohol content in the set of wine samples, record your results, and go home satisfied with a successful workday.

The next morning, when you come back to work, your colleague greets you with another set of wines to be measured. Easy job you think, since you already have your calibration curve. You noticed, however, that someone has removed your multiple reflection ATR unit from the spectrometer. You put it back in and proceed to optimize the optical alignment of the unit. This simply means that you adjust the mirrors in the ATR unit to maximize the overall signal on the spectrometer's detector.

Now everything is back to where it would have been had the unit not been removed from the spectrometer in the first place, so you proceed to make measurements and bring the results to your colleague. A few minutes later, your colleague comes to the lab and asks if you knew why today's results are about 5% higher than yesterday's. It turns out that your colleague has divided the original set of samples into two identical sets so that he can test the accuracy of the experimental method. It also turns out that he just happened to have read this book and thus became aware of the possible effect of the alignment of the

ATR unit on the effective number of reflections. He was also the one who removed the ATR unit from the spectrometer to compel you to realign it.

What has happened? By aligning the unit in the spectrometer, you optimize the signal on the detector, which means that you maximize the number of rays that make it through your ATR unit and on to the detector. However, in doing so, you may be settling onto a slightly different geometry than the one shown in Figure 15.8. This slightly different geometry may actually result in the same number of rays making it through the ATR element, hence yielding the same detector signal level, but the fraction of rays undergoing a different number of reflections may somewhat change. For instance, the hypothetical expression (Eq. 15.14) may now look something like

$$T = 0.05R^{10} + 0.45R^{9} + 0.35R^{8} + 0.15R^{7}. \quad (15.15)$$

Clearly, the ATR measurement governed by Equation 15.14 will produce a slightly different result from a measurement governed by Equation 15.15. Figure 15.10 illustrates the difference by showing the multireflection spectra of a hypothetical sample, one governed by Equation 15.14 and one by Equation 15.15. The small differences seen in Figure 15.10 explain the differences between the two measurements.

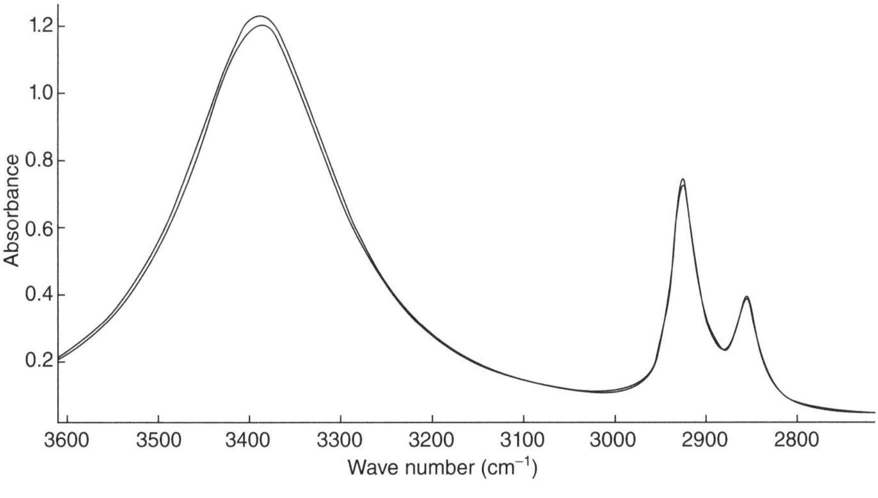

Figure 15.10 Multiple reflection ATR spectra of the same sample by the same multireflection ATR unit that was independently optically optimized for maximum energy before each measurement.

15.7 CHANGE IN THE REFRACTIVE INDEX OF THE SAMPLE DUE TO CONCENTRATION CHANGE

Let us briefly revisit the issue of a linear relationship between the concentration and the absorbance in ATR spectroscopy because we would like to spotlight the effect of change in the refractive index of the sample as a consequence of a change in the concentration of ingredients. Imagine two samples, one with a refractive index of 1.40 and one with a refractive index of 1.50. These refractive indices are typical for real-life samples. If we now calculate the spectra of a set of mixtures of the two materials from 0% to 100% in 10% steps, not only the concentration of the two components change but the refractive index of the mixture changes as well. The change in the refractive index changes the effective thickness of the sample as given in Equation 8.5. The absorbance is proportional to the product of concentration and effective thickness, and since the effective thickness is a function of concentration, there is a nonlinear response of absorbance to changes in concentration. The result is shown in Figure 15.11.

The lowest and the highest spectrum at each peak represent the two pure components. Keep in mind that the steps between the consecutive spectra are for equal changes in concentration. The peak near 1740 cm^{-1} shows increasing steps with the increasing concentration of the substance responsible for this particular absorption peak. The peak near 1630 cm^{-1} shows the opposite behavior; that is, the steps between the consecutive spectra decrease with the increasing concentration of the substance responsible for this particular absorption peak. Obviously, the two absorption peaks belong to two different components in the mixture.

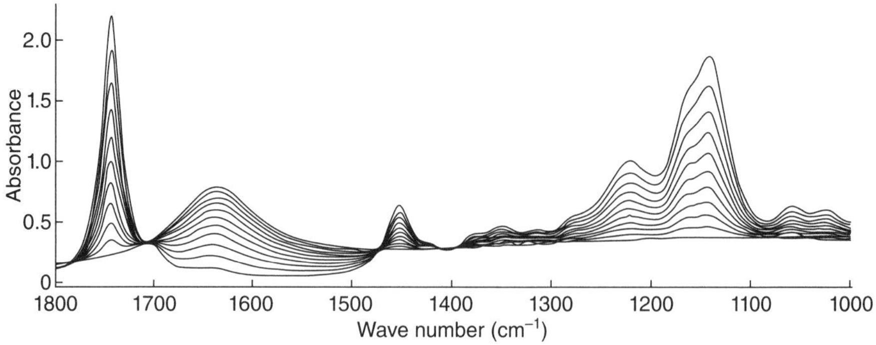

Figure 15.11 ATR spectra of two component mixtures.

The reason that the two peaks behave differently comes from what happens to the refractive index of the mixture. As the refractive index changes from the value of 1.40 to the value of 1.50, the effective thickness increases. Thus, the steps between the intensities of a peak that belongs to the material with the refractive index 1.50 grow with the increasing concentration of that material. Conversely, the steps between the intensities of a peak that belongs to the material with the refractive index 1.40 decrease with the increasing concentration of that material. This type of nonlinear response of absorbance to changes in concentration is present in ATR spectroscopy in addition to all other possible causes of nonlinearity that we have already discussed.

16 ATR Spectroscopy of Small Samples

16.1 BENEFITS OF ATTENUATED TOTAL REFLECTION (ATR) FOR MICROSAMPLING

One of the most useful variants of ATR spectroscopy is the technique known as ATR microsampling. The reasons for this usefulness are multiple and we will review some below.

Obviously, microsampling enables the analysis of very small samples. This characteristic of microsampling is crucial in those applications where samples to be analyzed are very small or are embedded within a larger matrix. For microsampling applications, the spectrometer's beam spot must be reduced in size. This is usually accomplished with the help of powered mirrors. In infrared (IR), lenses are avoided because of the chromatic aberration intrinsic to their function.

Another important feature of ATR spectroscopy that is very useful for microsampling is that ATR spectroscopy, unlike transmission spectroscopy, does not require the sample to be flattened into a thin film. ATR spectroscopy automatically sets its own "path length" through the sample.

The use of hemispherical ATR elements provides a further reduction of the spectrometer beam. For example, a reduction of the beam size by 6X using powered mirrors would reduce the original spectrometer's beam with a diameter in the focal point of 6 mm to a spot that is 1 mm in diameter. This reduction occurs at no expense to the beam's intensity.

However, in the process, the original beam spread of ±7° increases to ±47°. Obviously, the beam spread could hardly be increased much beyond ±47°. To analyze samples much smaller than 1 mm in diameter, a beam with a spot size much smaller than 1 mm in diameter is needed. So, a further beam reduction can only occur by cutting away the beam and consequently reducing the beam's intensity.

Internal Reflection and ATR Spectroscopy, First Edition. Milan Milosevic.

This is why the additional beam spot reduction provided by a hemispherical ATR element is so helpful for microsampling. For instance, using a germanium hemisphere, an additional beam reduction of 4X can be achieved without increasing the beam spread of the beam. This additional beam size reduction produces a beam with a spot size of 0.25 mm. Thus, all the intensity that was flowing in the original spectrometer's beam, which is 6 mm in diameter in its focal point, still flows through the spot size that is now 0.25 mm in diameter.

Another benefit that ATR spectroscopy provides to microsampling has to do with recording the spectra of samples whose size is on the order of the wavelength of light used for the measurement. The wave nature of light imposes restrictions on how small a spot can be formed by a particular wavelength of light. This restriction is called the diffraction limit. It is the same restriction that limits the resolution of optical microscopes. The rule of thumb is that light of a certain wavelength cannot form a spot that is much smaller than that wavelength. In the IR region, where ATR spectroscopy is mostly used, this amounts to a spot size limitation between 2.5 and 25.0 μm. The most information-rich region of an IR spectrum is typically in the longer wavelength range of the spectrum. This limits the spot size to about 10–20 μm.

A smaller sample can still be analyzed, but some of the light will "go around" the sample and will not interact with it. The light that interacts with the sample gets "diluted" by the light that does not interact with the sample. The end result is that this dilution impacts the linearity of the relationship between the measured absorbance and the absorption coefficient of the sample. In addition, for a sample embedded in a matrix, this go around light interacts with the matrix and the spectrum of the sample cannot be isolated from that of the matrix.

This diffraction effect that limits the spot size of the beam can be alleviated by the reduction of the wavelength size that occurs when light propagates through a material with the refractive index n. As we have seen, the speed of light c decreases in the medium of refractive index n to $c' = c/n$. Since the wavelength is determined by how far light travels during one full oscillation, this decrease in speed causes the wavelength to become n times smaller than it is in vacuum (or air). Again, in a high refractive index material such as germanium, this effect is huge. The wavelength reduction in germanium is four times, thereby significantly alleviating the restrictions posed by the diffraction limit. Note that the diffraction limit still applies, but in a high refractive index optical material, the wavelength is reduced, while the contact area between the sample and the ATR element remains the same.

16.2 CONTACT PROBLEM FOR SOLID SAMPLES

For irregular hard solid samples, there is an additional benefit that comes from a small sampling spot (often referred to as hot spot) that could be best understood as follows. Assume that the sample is a hard polymer sphere such as a Teflon ball. To get an ATR spectrum of the sphere, one needs to press the sphere against the sampling surface of the ATR element and locally deform it, so that the sphere and the ATR element are in direct physical contact over a circular spot of radius r_0. To avoid stray light, the size of the contact area has to be equal to, or larger than, the size of the hot spot.

The pressure distribution over the contact area has the maximum in the center of the contact circle and vanishes at the edges. The radius of the contact circle grows with the increased force on the sphere.

The mechanical problem of an elastic sphere in contact with a hard plane is well understood and is known as the *Hertz contact problem.* The expressions relating the radius of the contact circle, applied force, and maximum pressure (that occurs in the center of the contact circle) are

$$r_0 = \left[\frac{2}{3}FR\frac{1-v^2}{E}\right]^{\frac{1}{3}} \qquad (16.1)$$

and

$$p_o = \frac{3F}{2\pi r_0^2}, \qquad (16.2)$$

where R is the radius of the sphere; E and v are the Young modulus and the Poisson ratio of the material, respectively; F is the applied force on the sphere; and *po* is the pressure in the center of the contact circle.

Equations 16.1 and 16.2 can be combined to express p_o in terms of the radius of the contacting circle:

$$p_o = \frac{Er_0}{\pi R(1-v^2)}. \qquad (16.3)$$

In order to avoid damaging the surface of the ATR element, this pressure must be smaller than the modulus of rupture of the material that the ATR element is made from. As Equation 16.3 shows explicitly, a larger hot spot requires a proportionally larger contact pressure and, according to Equation 16.1, a much larger applied force.

A large applied force has several unwanted consequences. One is that the measurement is no longer nondestructive. A high contact force can damage or completely destroy the sample. Also, the sample morphology may change due to high contact pressures and the structural information thus obtained may be compromised. Independent of that, the mechanical support for the ATR element may flex in response to applied forces, affecting the optical alignment of the sampling accessory and resulting in unwanted baseline shifts and spurious spectral features due to the unequal optical paths between the sample and reference spectra, and so on. Hence, it is beneficial to reduce the hot spot size to as small as possible.

However, hot spot size reduction is optically coupled with the increase in the power of the converging/diverging beam. The increase in power of the incident beam means that the specificity in the angle of incidence is degraded. One can restore the definition in the angle of incidence by blocking the rays that fall outside the desired range for the angles of incidence, but that would be done at the expense of light intensity used in the experiment and hence at the expense to the signal/noise (S/N) ratio of the measurement.

17 Surface Plasma Waves

17.1 EXCITATION OF SURFACE PLASMA WAVES

Surface plasma waves are a topic somewhat tangential to our main subject of interest. However, we have encountered surface plasmons in our study of internal reflection of light on thin metal films deposited over a totally internally reflecting surface of an attenuated total reflection (ATR) element. What we observed in Figure 12.7 is that the reflectance of a thin film for a *p*-polarized beam drops near zero just after reaching total reflectivity at the critical angle. In the discussion associated with that figure, we focused on the curious fact that the reflectance is total at the critical angle. Now we focus on the plasmons. The refractive index of the ATR material was $n_1 = 2.42$ and that of the metal was $n_2 = 3 + 10i$. The material behind the metal film was air ($n_3 = 1$). Going back to Figure 11.7, we see that the internal reflection in this case is of the type (b). The metal–ATR element interface is subcritical, so the wave in the metal film is not evanescent. It is a true propagating wave.

Of course, a metal, having a huge absorption index, presents a formidable obstacle to a propagating electromagnetic wave, so we do not see fringes normally associated with a type (b) internal reflection. The propagating wave gets absorbed too quickly. Nevertheless, it is important to emphasize that the wave in the metal film is not an evanescent wave.

The refractive index that we used for metal has the real part larger than the refractive index of the ATR material. The critical angle, marked by the huge spike in reflectance at about 24°, is the critical angle for the ATR material–air interface. Note that the reflectance of the film, in observance of the Brewster angle, makes a valiant effort to dip again near 65°. This further confirms that the reflectance is not total.

The reflectance in Figure 17.1 exhibits a sharp and narrow absorption peak dropping down almost to zero at an angle near 25°. This peak

Internal Reflection and ATR Spectroscopy, First Edition. Milan Milosevic.

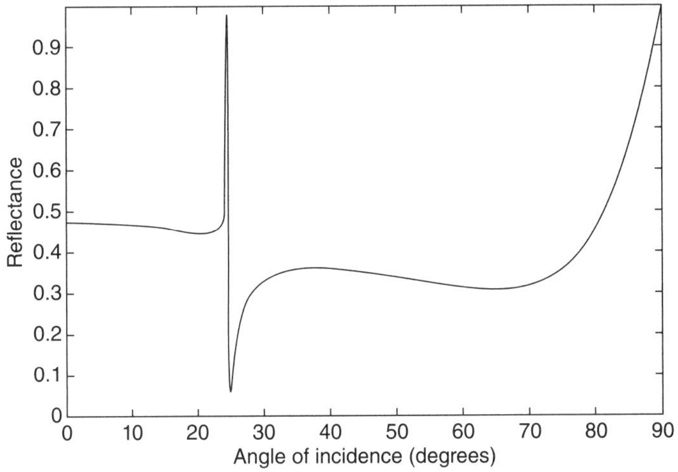

Figure 17.1 Internal reflectance of a thin metal film showing absorption due to the excitation of surface plasma waves.

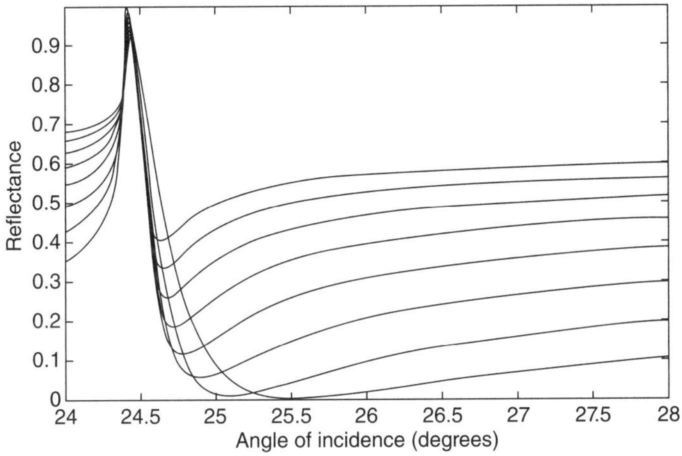

Figure 17.2 Effect of film thickness on reflectance.

is due to plasmon absorption. Let us examine how the various experimental parameters affect the plasmon peak position and strength.

17.2 EFFECT OF METAL FILM THICKNESS ON REFLECTANCE

Figure 17.2 shows how the film thickness of the metal film affects the angular dependence of reflectance. The reflectance curves shown represent film thicknesses from 30 Å (the curve with the lowest value at

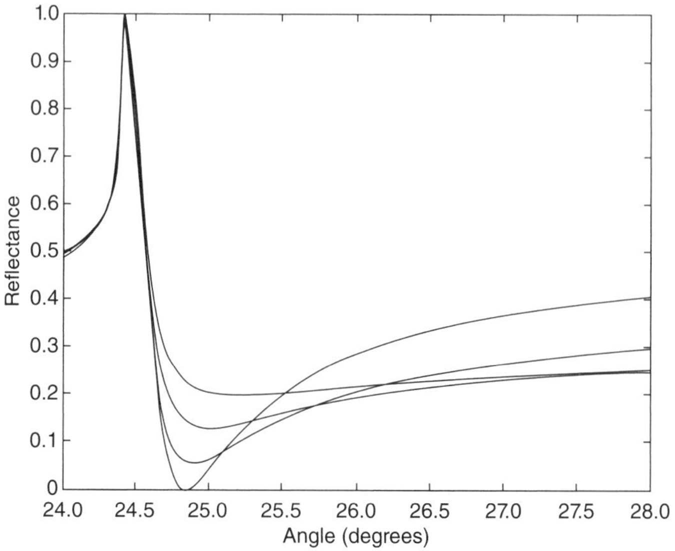

Figure 17.3 Effect of the real part of the refractive index on reflectance.

28°) to 100 Å in steps of 10 Å. We see that the thicker the film, the more pronounced is the plasmon absorption. Also, the peak position moves to lower angles as the thickness increases.

17.3 EFFECT OF THE REFRACTIVE INDEX OF METAL ON REFLECTANCE

Figure 17.3 shows the effect of change in the real part of the refractive index of metal. The film thickness is 50 Å, and the real part of the refractive index changes from 2 to 5, keeping the imaginary part constant at 10. The curve with the lowest value at the peak corresponds to the real part of the refractive index of 2, the next lowest for the value of 3, and so on.

17.4 EFFECT OF THE ABSORPTION INDEX OF METAL ON REFLECTANCE

Similarly, we can find the effect of change in the absorption index of a metal film on reflectance. Figure 17.4 shows the reflectance curves for the absorption index values of 10, 15, and 20 while keeping the real part

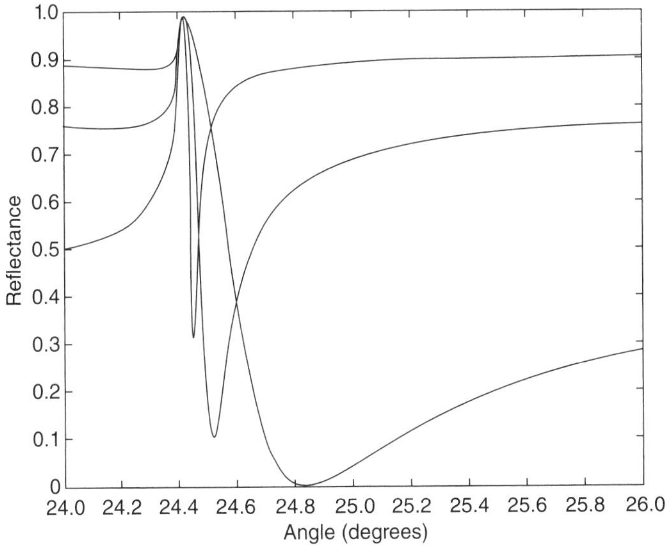

Figure 17.4 Effect of the absorption index of the metal on reflectance.

of the refractive index constant at 2. The effect of the absorption index is clearly very dramatic. The lowest curve is for the absorption index of metal of 10, the next lowest for 15, and the highest for 20. Note the dramatic sharpening of the absorption peak as the absorption index of metal increases. Interestingly, even though the peak becomes narrower with an increasing absorption index, the minimum value of the peak increases. Note that the absorption peak reaches down to zero reflectance for the metal absorption index of 10 and the angle of incidence just above 24.8°.

17.5 USE OF PLASMONS FOR DETECTING MINUTE CHANGES OF THE REFRACTIVE INDEX OF MATERIALS

The excitation of plasmons in thin metal films attracts growing interest as a very sensitive way of detecting the minute changes in the refractive index of the material behind metal. Thus, it is of some interest to see how a slight change in the refractive index of the third material changes the reflectance curve. For this purpose, let us assume that material 3 has a refractive index of 1.400. This will move the critical angle to a higher angle than it is for air. We assume that material 3 is nonabsorbing. A good real-life example would be that of alcohol in water, such as the case of wine that we discussed earlier.

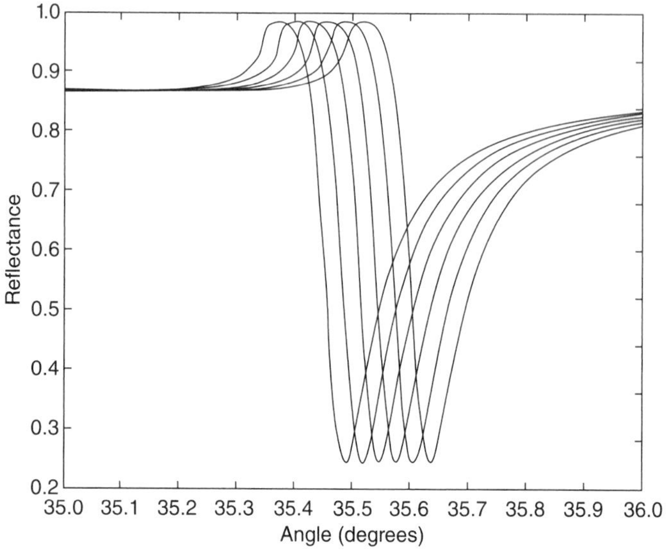

Figure 17.5 Changes in plasmon absorption in response to changes in the refractive index of the sample.

The ATR element is assumed to have the refractive index 2.42. The assumed refractive index of metal is $2 + 20i$. The thickness of the metal film was set to 50 Å. The wavelength of light is assumed to be 500 nm. Figure 17.5 shows the reflectance curves for the case when the contacting material (material 3) changes the refractive index from 1.400 to 1.405 in steps of 0.001. Note that the shift is fully resolved. Thus, the ultimate sensitivity of the technique appears to be much higher than third decimal place in the refractive index. It is also interesting that, unlike in the previous cases where changing a parameter affected not only the peak position but the general shape of the reflectance curve, the changes in the refractive index of the sample only shifts the peak position. The shape of the curve appears to be completely preserved. This fact can be very helpful in developing algorithms to use the entire curve (not just its value at a particular angle) to push the sensitivity of the measurement to the fourth or even fifth decimal place.

It is also of interest to see how the reflectance curves would look for a fixed angle of incidence, but using a full spectrum of wavelengths rather than a single wavelength in the variable angle measurements from Figure 17.6. Again, we set up the parameters as $n_1 = 2.420$, $n_2 = 2 + 20i$, $n_3 = 1.400$, the film thickness 50 Å, and the angle of incidence 35.5°. Notice that all three refractive indices were modeled as constant. This is not realistic since the refractive index always changes

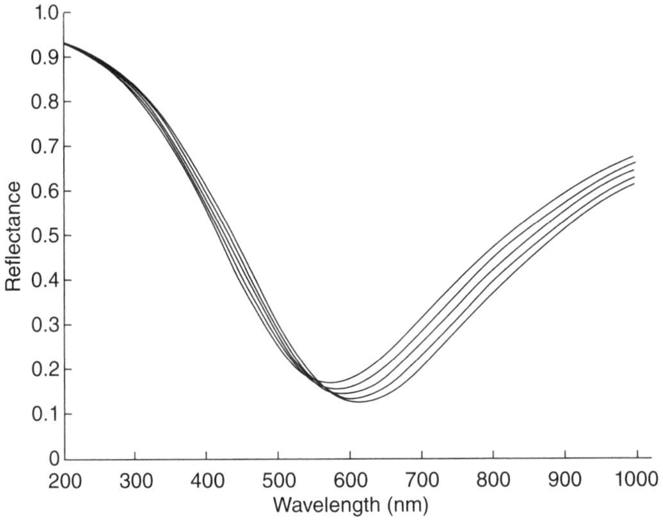

Figure 17.6 Changes in plasmon spectrum in response to changes in the refractive index of the sample.

with the wavelength, but it does not limit the validity of the simulation. The five values of the refractive index of the sample were 1.3998, 1.3999, 1.4000, 1.4001, and 1.4002. The results of these simulations are shown in Figure 17.6.

The spectrum with the strongest absorption peak is for the refractive index of 1.3998. The absorption peak shifts from 573 nm for the refractive index of sample of 1.4002 to 612 nm for the refractive index of 1.3998. That is 39 nm for a 0.0004 change in the refractive index or roughly 10 nm per 0.0001 of change in the refractive index. This is quite an exceptional sensitivity. Spectra are routinely collected with a resolution of 1 nm, so our results thus indicate the fifth decimal place sensitivity of the technique due to changes in the refractive index. This extraordinary sensitivity is behind the recent explosion in interest in surface plasma waves, both in the fundamental research community and in the industry as a means to construct very sensitive sensors for various industrial applications.

The above described potential applications of surface plasmons are based on the very high angular resolution of described measurements. As we can see from Figure 17.5, the significant changes in reflectance occur for angle of incidence changes significantly smaller than 0.1°. This degree of angular resolution is quite common for laser beams. The availability of tunable lasers makes these experiments with surface plasmons routine.

Plasmons are waves of charge density propagating on the surface of the film—thus electrical in nature.

The electromagnetic waves used to excite plasmons are optical phenomena. The described interactions between plasmons and light waves offer great promise for a direct interface between communication networks based on light signals traveling through optical fibers, and computers that are based on signals in the form of electrical currents traveling through copper wires.

17.6 USE OF PLASMONS FOR DETECTING MINUTE CHANGES OF THE ABSORPTION INDEX OF MATERIALS

Until now, we have considered only nonabsorbing samples. The question naturally arises as to what happens in the case of an absorbing sample. Figure 17.7 shows reflectance curves for the case when the sample is an extremely weak absorber. We keep all the parameters the same as for the spectra in Figure 17.6, except that we fix the real part of the refractive index of the sample to 1.400 and increase the absorption index of the sample in 0.0001 steps from zero to 0.0003. The spectrum with the strongest absorption peak in Figure 17.7 is, counterintuitively, the one for the zero value of the absorption index of the sample. The spectra with decreasing absorption correspond to the

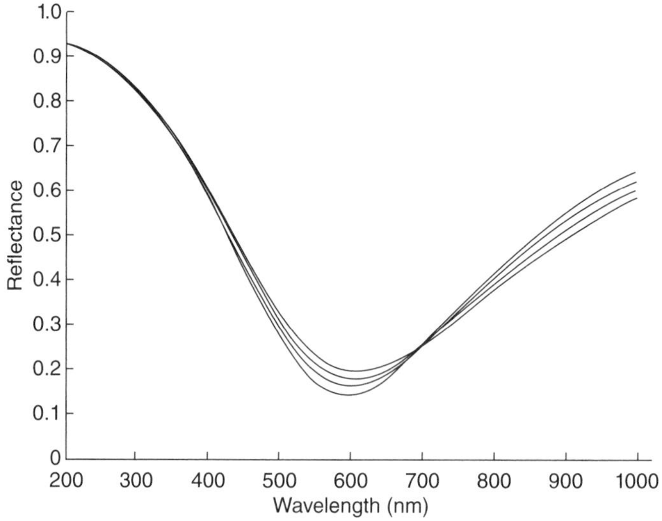

Figure 17.7 Changes in the plasmon spectrum in response to changes in the absorption index of the sample.

increasing value of the absorption index. The changes observed are of similar magnitude as what we saw for the same magnitude of changes in the real part of the refractive index (Fig. 17.6). This time, however, there is a curious "stationary" point near 700 nm. The absorbance at this wavelength is not changing with the changes in the absorption index of the sample. This indicates that, at least in principle, it is possible to separate the changes in the real part of the refractive index of the sample from the changes in the absorption index. The phenomenon of surface plasma waves has been known for a long time, but the renewed interest in the phenomenon may still uncover some unexpected aspects of it. It is important to realize that the entire phenomenon of surface plasma waves is already embedded into classical electromagnetic theory. We did not have to introduce any new mechanisms into our standard model to produce the above graphs. Everything that was modeled was done so based on expressions (Eq. 11.6). We did not even have to use any particular model for the refractive index of the metal film or of the sample. All we had to do is to provide the refractive indices, independent of any model, and to use standard optical theory as codified in expressions (Eq. 11.6). The plasmon phenomenon simply emerged out of the same equations that we used to calculate transmission and reflection spectra throughout this book. This again testifies to the richness of the world described by the deceptively simple expressions given by the classical electromagnetic theory.

18 Extraction of Optical Constants of Materials from Experiments

18.1 EXTRACTION OF OPTICAL CONSTANTS FROM MULTIPLE EXPERIMENTS

The optical constants of a material completely determine how light interacts with the material. Unlike spectroscopic observables, such as reflectance or transmittance, which incorporate dependence on geometric and experimental parameters like sample thickness or angle of incidence, respectively, optical constants are intrinsic properties of materials similar to specific gravity, or electrical conductivity. Thus, it would be desirable to extract optical constants from a spectroscopic measurement instead of reporting spectroscopic observables such as transmittance.

It is difficult to directly compare spectra of a particular material that have been recorded in two different experiments. Not only that the two samples have to be identical, but the spectroscopic techniques must also be identical, together with all the experimental parameters.

On the other hand, if the optical constants of a material are reported, it is irrelevant which technique was used and what were the experimental parameters. One could compare optical constants extracted from transmission measurements with those extracted from specular reflection or attenuated total reflection (ATR) measurements.

A spectral response of a material is described by two optical constants, $n(k)$ and $\kappa(k)$. Since there are two unknowns, we need two independent measurements of the spectroscopic observables of a material with all the experimental parameters known, to obtain the two equations necessary to extract the two unknowns. As we have seen before, the expressions relating optical constants to spectroscopic observables are quite complicated. Therefore, it is generally impossible to explicitly solve the system of two equations with two unknowns and

Internal Reflection and ATR Spectroscopy, First Edition. Milan Milosevic.

to provide direct analytical expressions for the calculation of optical constants. It is therefore necessary to resort to numerical methods. A method such as the following could be applied.

Let us denote the two spectral measurements obtained by changing the experimental parameter Q as

$$\begin{aligned} S_1 &= S(k, n, \kappa, Q_1) \\ S_2 &= S(k, n, \kappa, Q_2). \end{aligned} \tag{18.1}$$

The parameter Q may represent a single parameter or a group of parameters that change between the two measurements; k is the wave number, and n and κ are the optical constants.

The two experimental values corresponding to the expressions (Eq. 18.1) are $S_1^{\mathrm{exp}}, S_2^{\mathrm{exp}}$. We can never expect that the theoretical values will perfectly reproduce the measured values, but we can reasonably expect that the discrepancies between the measured and the calculated values are minimized for the correct choice of optical constants. In other words, for the correct optical constants, the quantity M, defined as

$$M(n, \kappa) = (S_1^{\mathrm{exp}} - S(k, n, \kappa, Q_1))^2 + (S_2^{\mathrm{exp}} - S(k, n, \kappa, Q_2))^2, \tag{18.2}$$

is minimized. This requirement allows for some noise in experimental values. Note that M is by definition always a positive quantity. Only if the choice of n and κ perfectly reproduces the measured values is M equal to zero. It is thus possible to numerically minimize M and thus find optical constants from experiment at a particular wave number k. This procedure must be repeated at every wave number in the range of interest. This may look like a long and complicated procedure, but it is not as bad as it looks. n and κ are smooth functions of wave number, and a known solution for n and κ at a particular wave number k is generally a good choice of a starting point for the iterations to minimize (Eq. 18.2) at a neighboring wave number. So the minimization would not take as long as may seem at first sight.

A good way to find the minimum of the sum (Eq. 18.2) is to start at a point n_0, κ_0 and to find the gradient of M at that point:

$$\nabla M = \left(\frac{\partial M}{\partial n}, \frac{\partial M}{\partial \kappa} \right). \tag{18.3}$$

The gradient of M is a vector that points in the direction in which function M grows the fastest. Thus, by advancing in the opposite direction,

$$n = n_0 - \frac{\partial M}{\partial n}\delta$$
$$\kappa = \kappa_0 - \frac{\partial M}{\partial \kappa}\delta, \tag{18.4}$$

we move through the $n - \kappa$ plane in the direction of the steepest descent of the function M. That way, we reach the point in the n-κ plane in which M has the lowest value.

Step δ is some small value such as 0.001 or so. It may be advantageous to initially choose a larger step to advance quickly toward the minimum of M and later to switch to ever finer steps to improve the resulting precision of the values of n and κ.

Note that we do not know if the procedure will lead us to the absolute minimum of M. The method is leading us to a minimum and the iterations could get stuck in a local minimum. A way out of it could be to randomly (within a realistic range) select a number of starting points in the n-κ plane and to let the procedure find local minima for each starting point independently. If all these starting points lead to the same minimum, chances are that this is the absolute minimum, but nothing is ever guaranteed.

Note again that this elaborate multipoint procedure is only needed for the first wave number in the spectrum. For all the subsequent wave numbers, a good starting point is the solution for the preceding wave number.

Finally, an important point that needs to be noted about the described procedure is that the quantity M in Equation 18.2 can include a much larger number of experimental spectra, not just two. The larger the number of experimental spectra included in the sum, the higher the expected accuracy of the results.

Note, however, that there must be at least two independent spectra to provide sufficient information to extract the two unknowns. For instance, the reflectance spectra at a 45° angle of incidence for s- and p-polarized incident light would not be sufficient since, as we have seen, $R_p = R_s^2$.

Let us further note that some spectroscopic techniques are more sensitive to n and some to κ. For instance, ATR and transmittance are generally more sensitive to κ, while specular reflectance is generally more sensitive to n. What we mean by higher sensitivity to κ is that a small change in κ causes a much larger change in the spectrum than an equally small change in n. Thus, it may be prudent to select for the above numerical procedure at least one spectrum more sensitive to n and at least one more sensitive to κ.

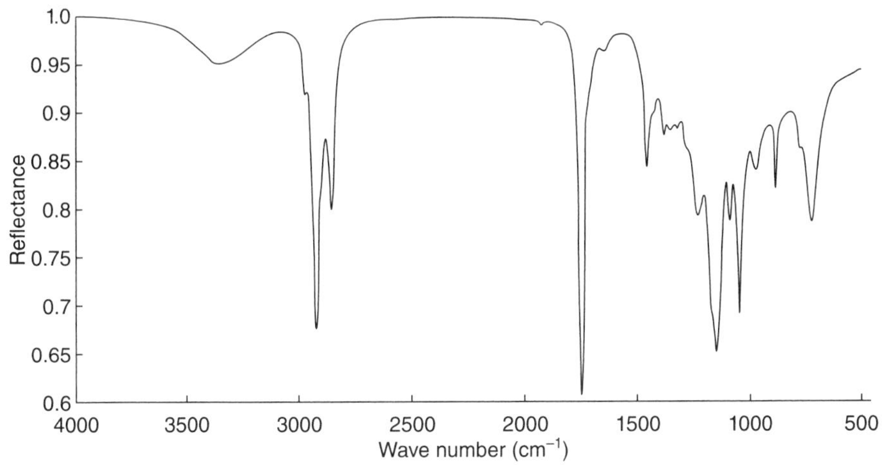

Figure 18.1 ATR spectrum of a sample, 45° angle of incidence, *p*-polarized light.

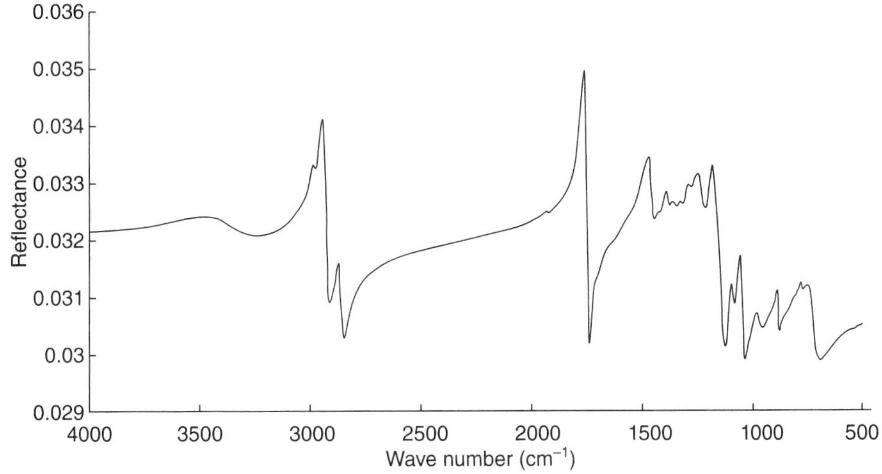

Figure 18.2 Subcritical internal reflection of the sample from Figure 18.1, 20° angle of incidence, *p*-polarized light.

For instance, a good choice would be to select the internal reflection spectrum just under the critical angle for total internal reflection and one just above. Figure 18.1 shows the ATR spectrum of a sample above the critical angle for total internal reflection, and Figure 18.2 shows the spectrum of the same sample, but for an angle of incidence below the critical.

Clearly, it is important to specify the polarization of the incident beam. The polarization of a typical spectrometer beam is generally not well-defined since mirrors and other optical components partially polarize the beam and also make the polarization of the beam somewhat dependent on the wavelength of light. So it is safest to use a polarizer and to select either the *s* or the *p* polarization. Figures 18.1 and 18.2 show spectra for the *p*-polarized beam. We could have just as well used the *s*-polarized beam.

The remarkably different appearance of the two spectra testifies to the different sensitivities of subcritical and supercritical internal reflection to the two optical constants.

If we consider the level of "wiggling" in the spectrum as indicative of its information content, then the two spectra shown in Figures 18.1 and 18.2 appear to carry the same amount of information about the sample. It somehow seems that the information contained in the ATR spectrum of Figure 18.1 is just replayed in Figure 18.2, through some precise but unknown "transformation." Loosely speaking, the spectrum of Figure 18.2 looks like a "derivative" of the spectrum from Figure 18.1. The fact that the two spectra look interconnected should come as no surprise since both spectra derive from the same complex refractive index.

The question then naturally arises: Is there some deeper connection between the two optical constants? Our harmonic oscillator model clearly connects the two as resulting from the same set of parameters. However, it turns out that a more formal connection can be drawn between the two optical constants.

18.2 KRAMERS–KRONIG RELATIONS

In most of the above, we have treated the two optical constants as two totally independent functions. It is true that we considered them as the real and the imaginary parts of the refractive index, but this seemed more as a part of a convenient notation. We did not impose any specific requirements on these functions. They could have been entirely arbitrary.

This "notational device" worked very nicely in making our expressions compact. It let us use the complex exponential function e^{ikx} as a solution for wave propagation and by making the refractive index a complex number, the imaginary part of the refractive index conspired beautifully with the complex exponential function to describe wave absorption in a medium. Actually, looking back over what we have done so far with the complex formalism of the electromagnetic theory, we should rightly be amazed with how little trouble it gave us. All our results came out in harmony with all the known laws of nature.

It would be, for instance, all too easy to imagine that some fairly involved calculations produced reflectance values larger than one. Or that we ran into another similarly unnatural result. But we did not, despite pushing the formalism into some pretty far out extremes, such as treating a metal as just another optical material.

This synchrony of mathematics and physics seems almost miraculous. After all, higher mathematics is derived from man-made axioms using pure logic. True, these rules were inspired by the observations and attempts to describe the physical world, but humans have made uncounted mistakes in the process of trying to understand and describe the physical world. So, the "unreasonable effectiveness" of mathematics as used in the physical sciences is at least puzzling.

We have considered the complex refractive index $n_c(k)$ of a medium as a function of wave number, where we have used not the wave number in the medium but the wave number of light in vacuum. The wave number in the medium is of course the wave number in vacuum multiplied by the refractive index of the medium, which is often a complex number. Thus, it could be considered that, in general, the wave number is a complex number and that the complex refractive index is thus a complex function of a complex wave number. Although we may prefer to work with a complex refractive index and to keep the wave number in our expressions real, it comes natural that the wave number could be extended to complex numbers.

A special kind of complex function, which encompasses most of the functions used in the physical sciences, is called an analytical function. This function has a remarkable property encapsulated by the so-called Cauchy integral:

$$f(z) = \frac{1}{2\pi i} \oint \frac{f(\zeta)d\zeta}{\zeta - z}. \tag{18.5}$$

A value of a function in a complex point z can be calculated from the values of the same function on some closed path encircling that point. In the case of a physical response function f, the arguments based on causality (i.e., cause precedes effect) imply that the function f is analytical in the upper complex plane. First, we select z to be on the real axis, that is, $z = x$. We can also choose the path of integration for the integral to go from $-\infty$ to ∞ along the real axis and then back along the infinite semicircle in the upper complex plane. If no singularities are encircled by the path of integration, the integral is zero. If $f(\zeta) \to 0$ as $|\zeta| \to 0$, the integral along the large semicircle vanishes. As the variable of integration ζ advances along the real axis $\zeta = x'$, where x' is a real number, it encounters a singularity at $x' = x$. We can avoid including this singularity into the area encircled by the path of integration by following a small semicircle around the singularity and then allowing the radius of that semicircle to go to zero. The contribution to the integral along this small semicircle is easily calculated:

$$\oint \frac{f(\zeta)d\zeta}{\zeta - z} = -\int_0^{\pi} \frac{f(x+\varepsilon e^{i\varphi})}{\varepsilon e^{i\varphi}} i\varepsilon e^{i\varphi} d\varphi = -i\pi f(x), \tag{18.6}$$

where we have substituted $\zeta = x + \varepsilon e^{i\varphi}$, let $\varepsilon \to 0$ at the end of the calculation, and we also reversed the limits of integration from $\pi \to 0$ to $0 \to \pi$, which accounts for the minus sign in front of the integral.

The original integral is zero since there are no singularities inside the area circumscribed by the path of integration. Therefore, the integral along the real axis plus the integral around the singularity at x adds to zero, or

$$P\int_{-\infty}^{\infty} \frac{f(x')dx'}{x'-x} = i\pi f(x).$$

The letter P in front of the integral stands for the principal value of the integral, which means that in evaluating the integral, we have to bypass the singularity at x by following the above procedure with a vanishing semicircle. By separating real and imaginary parts in the above integral, we find

$$\begin{aligned} Ref(x) &= \frac{1}{\pi} P\int_{-\infty}^{\infty} \frac{Imf(x')dx'}{x'-x} \\ Imf(x) &= -\frac{1}{\pi} P\int_{-\infty}^{\infty} \frac{Ref(x')dx'}{x'-x}. \end{aligned} \tag{18.7}$$

The relations (Eq. 18.7) express the real part of a function in terms of its imaginary part and vice versa. Note that here we used the assumption that both integrals in Equation 18.7 are finite. This assumption about $f(x)$ is critical to the validity of Equation 18.7. These integrals can only be finite if $Ref(x), Imf(x) \to 0$ when $x \to \pm\infty$. Clearly, the integrals cannot be finite otherwise. That becomes a problem if we try to apply Equation 18.7 to $n_c(k)$ since we know that the real part of n_c is almost never less than 1. However, if we use $n_c(k) - n(\infty)$ instead, the problem disappears and we can write

$$\begin{aligned} n(k) &= n(\infty) + \frac{1}{\pi} P \int_{-\infty}^{\infty} \frac{\kappa(k')dk'}{k'-k} \\ \kappa(k) &= -\frac{1}{\pi} P \int_{-\infty}^{\infty} \frac{(n(k') - n(\infty))dk'}{k'-k}. \end{aligned} \tag{18.8}$$

The above equations are the famous Kramers–Kronig dispersion relations. They can still make us a bit uneasy since they clearly employ negative wave numbers (or frequencies). There is no physical meaning to negative frequencies. It is, however, possible to infer, based on the principle of causality, that $n(-k) = n(k)$ and $\kappa(-k) = -\kappa(k)$. This is just an extension of the definitions of n and κ to negative frequencies that is consistent with the notion that an event could be influenced only by the events preceding it. If these symmetries are taken into account, the above equations could be expressed in terms of positive frequencies only.

The above expressions, at least in principle, enable us to calculate one optical constant from another. However, neither of the optical constants can be measured directly in an experiment, so in practice, the Kramers–Kronig equations cannot be used in the form given in Equation 18.8.

We now turn to the case of normal incidence reflectance that enables us to obtain both optical constants from a single normal incidence reflectance spectrum. In what follows, we will accomplish something that has a flavor of mathematical impossibility. We will use one experimental spectrum—a normal incidence front surface reflectance of a material—and extract from it both optical constants of the material. The mathematical procedure appears as if we succeeded in solving one equation with two unknowns. Mathematically, that is a big no-no, but we will see that no magic takes place in the calculations.

18.3 KRAMERS–KRONIG EQUATIONS FOR NORMAL INCIDENCE REFLECTANCE

We have found before that, for normal incidence, the reflectance amplitude coefficients for the two polarizations are equal.

The expression for the normal incidence reflectance coefficient takes a particularly simple form (Eq. 4.17a):

$$r = \frac{n_c - 1}{n_c + 1}.$$

The value of r is generally a complex number which we can rewrite using the polar form as

$$r = |r| e^{i\psi}. \tag{18.9}$$

Since the reflectance is $R = |r^2|$, we can rewrite Equation 18.9 as

$$r = \sqrt{R} e^{i\psi}. \tag{18.10}$$

By taking a logarithm of Equation 18.10, we find

$$\ln(r) = \frac{1}{2} \ln(R) + i\psi. \tag{18.11}$$

Notice that the logarithm of the reflectance amplitude coefficient is just an elementary function of the refractive index, so if the refractive index is an analytical function, any analytical function of the refractive index is then necessarily also an analytical function. Therefore, ln(r) from Equation 18.11 satisfies the same Kramers–Kronig relations (Eq. 18.8) as $n(k)$.

We can thus connect the phase of the reflectance amplitude coefficient with the logarithm of the measured reflectance through Kramers–Kronig relations as follows:

$$\psi(k) = -\frac{1}{2\pi} P \int_{-\infty}^{\infty} \frac{\ln R(k')dk'}{k' - k} = -\frac{k}{\pi} P \int_{0}^{\infty} \frac{\ln R(k')dk'}{k'^2 - k^2}, \tag{18.12}$$

where the second step follows from the symmetry property $R(k) = R(-k)$.

The above result is interesting but not necessarily useful, since we hardly ever need to know the phase of a reflectance amplitude coefficient. Anyway, we see that if we measured $R(k)$, we could, at least in principle, calculate $\psi(k)$. The magic of the Kramers–Kronig formalism emerges when we realize that we can analytically express $n(k)$ and $\kappa(k)$ in terms of $R(k)$ and $\psi(k)$. To see how, let us separate $r(k)$ into real and imaginary parts:

$$r = \frac{n^2 - 1 + \kappa^2}{(n+1)^2 + \kappa^2} + i \frac{2n\kappa}{(n+1)^2 + \kappa^2}. \tag{18.13}$$

By combining Equation 18.10 and 18.13, we arrive at

$$Re(r) = \sqrt{R}\cos\psi = \frac{n^2 - 1 + \kappa^2}{(n+1)^2 + \kappa^2}$$
$$Im(r) = \sqrt{R}\sin\psi = \frac{2n\kappa}{(n+1)^2 + \kappa^2}. \quad (18.14)$$

Now we can express $n(k)$ and $\kappa(k)$ in terms of $R(k)$ and $\psi(k)$ as follows:

$$n = \frac{1 - R}{1 + R - 2\sqrt{R}\cos\psi}$$
$$\kappa = \frac{2\sqrt{R}\sin\psi}{1 + R - 2\sqrt{R}\cos\psi}. \quad (18.15)$$

Notice what has occurred. We measured a single spectroscopic observable, $R(k)$, and have managed to extract two unknowns, $n(k)$ and $\kappa(k)$. It does have the flavor of solving one equation with two unknowns.

The formalism just outlined clearly offers a simple method to systematically extract optical constants from a straightforward experimental measurement. Moreover, we could use the expressions in Equation 18.15 even if the angle of incidence is not exactly normal but close to normal. The beauty of normal (or at least near normal) incidence is that it does not require the use of a polarizer so the entire spectrometer's beam is available for the experiment. This maximizes the signal/noise (S/N) of the measurement.

The fact that the optical constants can be explicitly analytically expressed in terms of the measured quantities is not really that important for the practical use of the Kramers–Kronig formalism. The calculation indicated in Equation 18.12 is only doable by a direct numerical evaluation, and clearly, in practice, calculations equivalent to Equation 18.15 can just as well be done numerically. Therefore, one could use the reflectance or even the transmittance of a sample at any angle of incidence and for either polarization to extract n and κ using the Kramers–Kronig formalism. It was also found that the use of the so-called fast Fourier transform (FFT) algorithm enormously speeds up the evaluation of the expressions (Eq. 18.12).

The Kramers–Kronig relations, however, require the reflectance to be known for all frequencies of light. No spectrometer could ever provide a complete spectral range. A typical Fourier transform infrared spectrometer usually covers the range from about 5000 to 400 cm^{-1}

(2–25 μm). Since a spectrometer covers the spectral range from k_1 to k_2, the integral (Eq. 18.12) can be separated into three terms:

$$\psi(k) = -\frac{k}{\pi} P \int_0^{k_1} \frac{\ln R(k')dk'}{k'^2 - k^2} - \frac{k}{\pi} P \int_{k_1}^{k_2} \frac{\ln R(k')dk'}{k'^2 - k^2} - \frac{k}{\pi} P \int_{k_2}^{\infty} \frac{\ln R(k')dk'}{k'^2 - k^2}. \quad (18.16)$$

Only the reflectance that appears in the middle term in Equation 18.16 is known from the experiment. Thus, some sort of extrapolation has to be done for the first and third terms. Even not doing any extrapolation for the reflectance outside the measured spectral region in effect amounts to making a specific extrapolation ($R = 1$). The size of the error depends on how good is the extrapolation that was used.

A reasonable check on the calculated optical constants is to use these constants to calculate from them the measured spectrum that was used to extract the optical constants. A comparison between thus calculated and the measured spectra can be made, and the degree of agreement can serve to gauge the expected error in the calculated optical constants.

A number of different forms of Kramers–Kronig equations have been derived that reduce the dependence of the integral on the values of the measured reflectance $R(k')$ far from the calculation point k and thus make the calculated optical constants less sensitive to the particular interpolation used. For instance, if we integrate Equation 18.12 by parts, we get

$$\psi(k) = -\frac{1}{2\pi} P \int_0^{\infty} \ln\left|\frac{k' + k}{k' - k}\right| \frac{d\ln R(k')}{dk'} dk'. \quad (18.17)$$

The benefit of the expression in Equation 18.17 over the expression in Equation 18.12 is that the contributions to the integral are narrowed to the portions of spectral range within which the reflectance changes. The spectral regions where the reflectance is constant do not contribute. Since we have seen that the reflectance undergoes rapid changes only within absorption bands, the integral (Eq. 18.17) effectively breaks down into a sum of integrals over the relatively narrow spectral ranges around the absorption peaks.

In addition, the optical constants being a universal property of the material examined and thus enabling easy comparison of measurements that were done using different spectroscopic techniques, also enable one to calculate how the spectra of the material would look if they were acquired by different spectroscopic techniques and/or using a different set of the experimental parameters. In other words, a particular feature of a sample could be easier to detect by a particular spectroscopic technique, so the search for the optimal measurement technique and for the optimal experimental parameters could be done without physically performing the different experiments.

19 ATR Spectroscopy of Powders

19.1 PROPAGATION OF LIGHT THROUGH INHOMOGENEOUS MEDIA

Up to now, we have assumed that the sample is uniform and isotropic, so the evanescent wave that is induced in the sample during supercritical internal reflection exists in an optically smooth medium. However, if a powdered sample is placed on an attenuated total reflection (ATR) element, the evanescent wave no longer encounters a smooth optical medium. Individual grains with air gaps between them present a number of reflective optical interfaces.

So what happens with the evanescent wave in a powdered medium? We should note that the individual grains of powder are solid material, but that the size of individual grains is comparable to the wavelengths of light impeding the formation of propagating waves within individual grains. Therefore, light does not propagate through a powdered medium in a straight line. Light propagation through a powdered sample is more similar to diffusion.

Let us first consider nonabsorbing powders. A good real-life example is that of snow. Water is completely transparent to visible light, as is ice. However, snow has quite a different appearance from water or ice. Snow is white, while water and ice are transparent.

Snow is formed by individual snowflakes sticking together, but because of their shapes, they do not fill the space completely, leaving tiny air gaps in the material. Light incident on the surface of snow partially reflects from the front surface and partially enters into the inhomogeneous material. Numerous tiny air gaps fill the medium with interfaces that serve as light scatterers distributed evenly and densely throughout the material. These scattering centers scatter the incident light in all directions, creating a random walk-like propagation of light through the medium. Some of the light is randomly scattered through

Internal Reflection and ATR Spectroscopy, First Edition. Milan Milosevic.

the material and ends up back at the surface where it is reemitted back out of the material.

A nice illustration of this can be seen on a winter night by aiming a laser pointer onto the surface of the snow. Light enters the snow at a point possibly several millimeters in size and generates the glow of reemitted light a few dozen centimeters in size.

Obviously, light propagates through powder differently than through a uniform homogeneous medium. This propagation is characterized by the absence of a specific light path. The direction of propagation randomly changes after traveling very short distances through the material. Therefore, the main characteristic of light propagation through a powdered material is the incessant scattering at virtually any point along the path and a diffusion-like spreading through the material.

If at some wavelengths of light the material absorbs, the fraction of incident light reemitted at those wavelengths is reduced with respect to those wavelengths for which the material does not absorb.

19.2 SPECTROSCOPIC ANALYSIS OF POWDERED SAMPLES

How does the evanescent wave propagate through an inhomogeneous material? Does it scatter, or does it propagate without being scattered? The first reaction would be that there is no apparent reason for the evanescent wave not to scatter. However, as we have seen when we studied it, the evanescent wave is a special kind of a wave that is bound to the surface. This binding mechanism manifests itself through the fact that the perpendicular component of the wave vector of the evanescent wave is imaginary.

If the wave scattered and became decoupled from the interface, the wave vector of the wave would have to become real. The scattered wave would be free to drift away from the interface and never to return back. That means that, for a nonabsorbing material, the supercritical internal reflection of the powder would become less than total.

However, it is an empirical fact that the supercritical internal reflectance of a nonabsorbing powder is total. No photons drift away from the interface. That implies that the evanescent wave is not scattered in the powdered sample. This is a surprising conclusion and we will return to this later.

The spectroscopic analysis of powders has always required special sample preparation. Traditionally, a powdered sample is mixed with a nonabsorbing powder such as KBr and is pressed into a pellet. The high pressure of the procedure forces the two powders to fuse into a clear disk that can then be analyzed by standard transmission spectroscopy. This procedure can be used to analyze virtually any solid sample. However, the procedure is laborious and requires a degree of skill that is more akin to art than science.

A spectroscopic technique that was embraced as a less laborious substitute for pellet making is called diffuse reflection spectroscopy. In this procedure, a sample is again ground and diluted in a nonabsorbing power such as KBr, but instead of pressing a pellet, the powdered sample is analyzed by reflection spectroscopy. Reflected (reemitted) light is characterized by transmission like dips at the wavelengths at which the sample absorbs. The benefit of this technique is that it does not require the skill and effort to produce pellets, so more samples can be analyzed per unit of time. Still, sample preparation is required and, while not impossible, sample recovery after the analysis would be quite involved. Also, quantitative measurements require careful mixing ratios of the sample and the diluting powder, repeatable compression of the resulting mixture for different samples, and so on.

Since light propagates through powder differently than through a homogeneous sample, the absorbance transform used to linearize transmission spectra is not appropriate for diffuse reflection spectra. A different linearizing transform, called the Kubelka–Munk transform, was instead developed for diffuse reflection spectroscopy. However, in order that the Kubelka–Munk transform is applicable to diffuse reflection spectra, the experimental arrangement of the diffuse reflection measurement must satisfy additional requirements assumed in deriving the Kubelka–Munk transform.

The use of ATR spectroscopy for powders proved to be extremely effective. No sample preparation is required. The sample is placed neat onto the ATR element, pressure is applied to compress it into good contact with the element, the spectrum is collected and the interpretation of the spectra is the same as for other types of samples.

Sample recovery, if desired, is straightforward. Much more samples per unit time can be analyzed so the use of ATR for powder analysis became common. This trend was further fueled by the introduction of ATR elements made from diamond. One of the downsides of using ATR for powder analysis is the possibility of the powder grains scratching the surface of the ATR element, especially so if the powder is a hard material. Although silicon ATR elements proved fairly scratch resistant, the use of diamond ATR elements completely muted that concern.

To analyze a sample by ATR, the sample has to be brought into intimate contact with the ATR element. The evanescent wave extends just a short distance above the surface of the ATR element—typically on the order of the wavelength of light used for the measurement. The sample needs to fill the small volume above the interface that is probed by the evanescent wave. If only a fraction of that volume is filled by the sample, the absorption experienced by the evanescent wave is smaller than if the volume is filled completely.

With solid samples, the usual approach is to use a pressure applicator to physically squeeze the sample against the ATR surface. With powders, the pressure applicator compactifies the sample by pressing the individual grains as close as possible. The grains are solid and usually irregularly shaped so they cannot be pressed together more than to the point where they touch. That inevitably leaves tiny empty spaces within the powder.

19.3 PARTICLE SIZE AND ABSORBANCE OF POWDERS

To see how particle size influences the ATR spectra of powders, let us consider an idealized case of a powder consisting of uniform spheres of radius R. We assume that the spheres are packed into the densest possible configuration (i.e., face-centered cubic structure). The situation is illustrated in Figure 19.1. From geometry it follows that the spheres are arranged into layers separated by the distance $H \approx 1.633R$. The adjacent layers are shifted so that a sphere always sits on the three spheres below it.

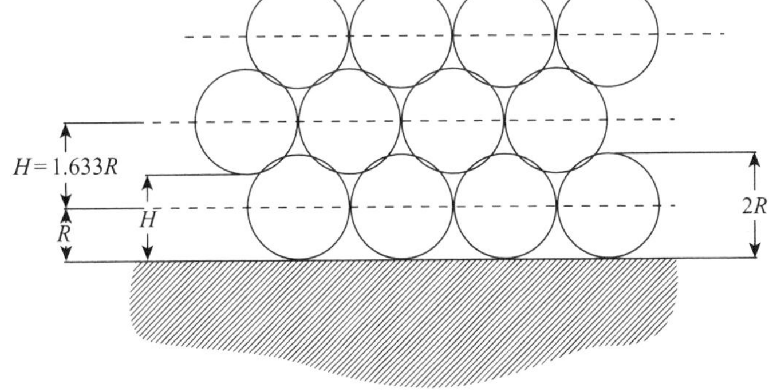

Figure 19.1 Idealized power consisting of uniform spherical grains packed on an ATR element.

It is obvious from Figure 19.1 that there is much more empty space near the surface of the ATR crystal than further away from it. As the intensity of the evanescent wave decays exponentially with the distance from the surface of the ATR element, it follows that the fraction of the volume occupied by the sample is smallest at the interface where the evanescent wave is strongest.

If we imagine a plane parallel to, and at a distance z, from the surface of the ATR element, this plane intersects spheres from either one layer or from two adjacent layers. The fraction of the surface area of that plane occupied by the spheres can be calculated from geometrical considerations. The result is shown in Figure 19.2. As we can see, the fraction of the surface occupied by the spheres varies periodically around the value of 74% as expected for a face-centered cubic structure.

As the typical grain size of fine powders is on the order of several microns and as the typical penetration depth of an evanescent wave in the infrared spectral region is also around several microns, the relevant portion of the graph in Figure 19.2 is up to several sphere radii.

The exact shape of the graph is also a consequence of the particular model used, but it would be reasonable to expect that, for a real powder, the curve takes off from zero, comes to saturation at around half of the average grain size, and remains constant thereafter.

Thus, one can replace the exact but artificial volume fraction function with a smoothed out version. The precise analytical expression for the

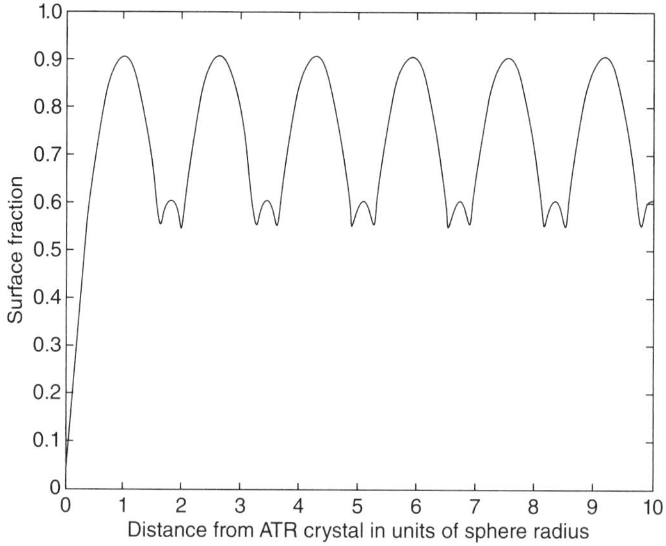

Figure 19.2 The fraction of the surface of a plane at a distance z from the surface of the ATR crystal occupied by spheres.

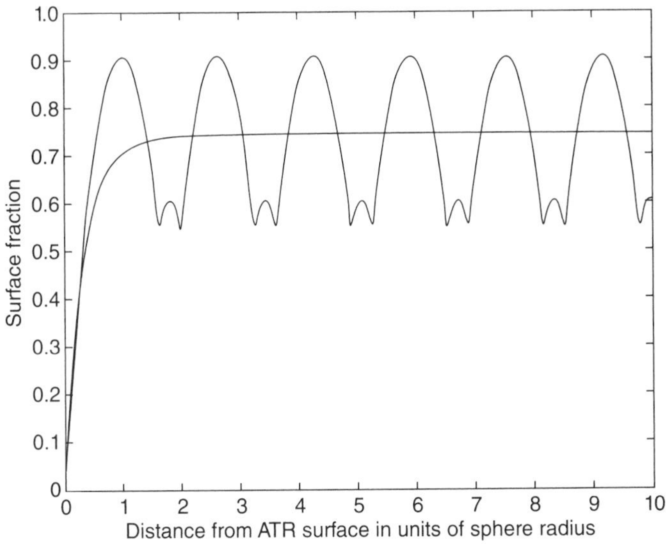

Figure 19.3 Actual and smoothed surface filling fractions for model powder consisting of spherical grains.

surface fraction function is not that important as long as it satisfies the general description given above. We may as well choose the form that is convenient for analytical manipulations:

$$f(z) = 0.7405 \cdot \left(1 - e^{-\frac{3z}{R}}\right). \tag{19.1}$$

The comparison between the real and smoothed volume filling fraction is shown in Figure 19.3. For a real powder, the grains will be packed randomly so the smoothed version of the filling fraction is more realistic. The smoothed function can be used to calculate how the absorbance of a powder is affected by grain size.

The absorbance of the evanescent wave in powder is found by following the steps from Equation 10.2. The relevant integral is

$$I = \int_0^\infty e^{-\frac{2z}{d_p}} \left(1 - e^{-\frac{3z}{R}}\right) dz = \frac{d_p}{2} \frac{1}{1 + \frac{2}{3}\frac{R}{d_p}}. \tag{19.2}$$

The first factor under the integral in Equation 19.2 describes the decay of the intensity of the evanescent wave with depth into the sample, and the factor in parentheses is the surface fraction occupied by the sample

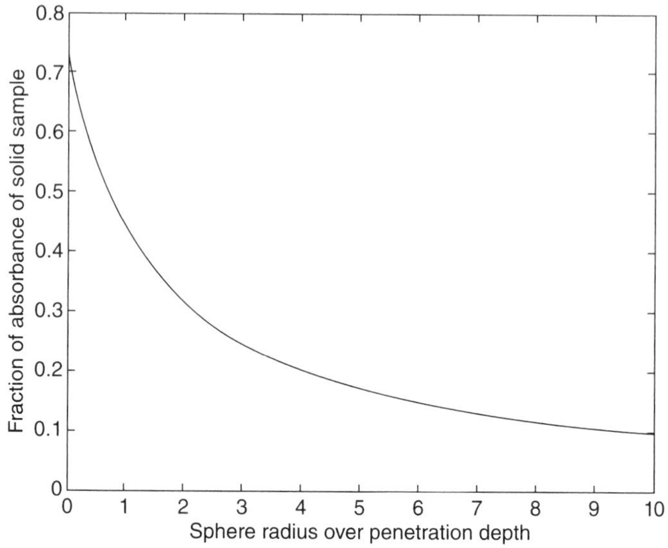

Figure 19.4 The absorbance of a powder as a function of the ratio of sphere radius and penetration depth.

at depth z. As grain size decreases, the absorbance approaches 74% of the value it would have if the sample was just a smooth solid without air gaps between grains. This implies that, by grinding the sample finer and finer, the absorbance of the sample grows stronger and stronger as shown in Figure 19.4, but no matter how small the grain size, the absorbance can never reach that of bulk material. It can only reach 74% of the bulk absorbance since the volume filling factor is independent of grain size. A tighter volume packaging is possible if the grains are not all the same size. Then, smaller grains can fill the space in the gaps between larger grains.

19.4 PROPAGATION OF EVANESCENT WAVE IN POWDERED MEDIA

We have already mentioned that in supercritical internal reflection spectroscopy of a powdered sample, an evanescent wave forms in an inhomogeneous optical medium. A powder is a medium rich with grain surfaces, each a strong source of scattering for a propagating electromagnetic wave. These surfaces optically represent interfaces between air and the solid material that constitutes powder. They reflect ordinary light with the effect that light propagates through powdered media not in a straight line but along a randomly zigzagging path.

The question that we are concerned with is how does the evanescent wave propagate through this highly scattering medium? What happens at the grain boundaries?

Can the evanescent wave scatter and become uncoupled from the internally reflecting surface and continue to propagate through the powder as a regular wave? Those are fundamental questions and are crucial to understanding the ATR spectroscopy of powders.

Let us start by reexamining the boundary conditions for the fields at the interface of two media. In deriving Fresnel equation for the *s*-polarized beam, we found (Eq. 4.6)

$$E_{in} + E_r - E_t = 0$$

$$n_1(E_{in} - E_r)\cos\theta - n_2 E_t \cos\varphi = 0.$$

We solved (Eq. 4.6) for E_r and E_t in terms of E_{in}. This produced very useful results but has also implied that the reflected wave is generated by the interface. The reflected wave is, of course, created by the induced dipole moments of the molecules of the medium, all of which are located behind the interface. These dipole moments were induced by the electric field of the transmitted wave propagating inside the medium, not by the external incident wave.

It was just the convenience of the mathematical formalism that led us to define reflection and transmission amplitude coefficients. We can just as well eliminate the incident field from the above equations and express the reflected wave in terms of the transmitted wave as

$$\frac{E_r}{E_t} = \frac{n_1 \cos\theta - \sqrt{n_2^2 - n_1^2 \sin^2\theta}}{2n_1 \cos\theta}. \quad (19.3)$$

Similarly, for the *p*-polarized light, we use Equation 4.9,

$$(E_{in} - E_r)\cos\theta - E_t \cos\varphi = 0$$

$$n_1(E_{in} + E_r) - n_2 E_t = 0,$$

to get

$$\frac{E_r}{E_t} = \frac{n_2^2 \cos\theta - n_1\sqrt{n_2^2 - n_1^2 \sin^2\theta}}{2n_1 n_2 \cos\theta}. \quad (19.4)$$

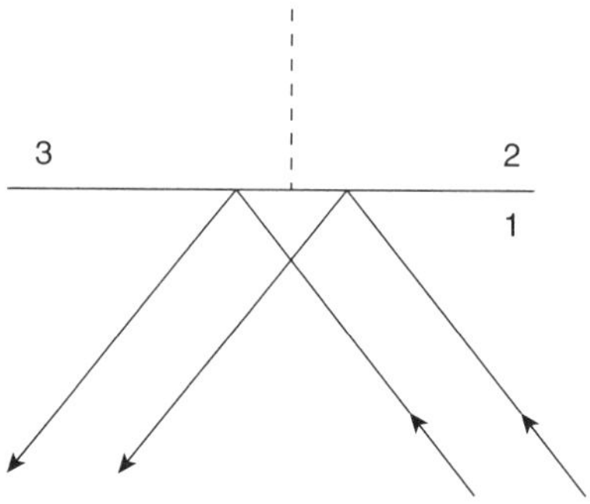

Figure 19.5 Light incident at the interface with two media.

The expressions (Eqs. 19.3 and 19.4) are, of course, totally useless from the practical point of view. However, they are instructive in terms of what causes what. These expressions are also helpful in the case of supercritical internal reflection in the sense that they reveal that it is an evanescent wave that generates a reflected wave. If an evanescent wave is not absorbed by the medium in which it propagates, the reflected wave carries off the same energy flow that was carried in by the incident wave. However, if the evanescent wave is absorbed by the medium in which it is induced, a weakened evanescent wave creates a weakened reflected wave. This picture explains ATR spectroscopy better than the concept that the reflected wave is somehow generated by the interface. After all, how would the reflected wave even know if the medium behind the interface absorbs?

The next situation that we need to carefully examine is shown in Figure 19.5. We assume that the refractive index of the ATR element n_1 is larger than n_2 and n_3 and that the reflection at the entire interface is supercritical. Materials 2 and 3 are divided by a boundary (shown as a dashed line in Fig. 19.5) perpendicular to the reflecting surface. Both sets of the incident rays, those that impinge on the 1–2 interface and those that impinge on the 1–3 interface, induce evanescent waves above the interface. Following Equation 5.3, we can write the expressions for the two evanescent waves as follows:

$$E_{12}^{\nu}(\boldsymbol{r},t)=t_{12}^{\nu}E_0^{\nu}e^{-i\omega t}e^{2\pi i n_1 kx\sin\theta}e^{-2\pi kz\sqrt{n_1^2\sin^2\theta-n_2^2}}$$
$$E_{13}^{\nu}(\boldsymbol{r},t)=t_{13}^{\nu}E_0^{\nu}e^{-i\omega t}e^{2\pi i n_1 kx\sin\theta}e^{-2\pi kz\sqrt{n_1^2\sin^2\theta-n_3^2}}. \quad (19.5)$$

The question arises: What happens with these evanescent waves at the boundary of media 2 and 3? A difficulty stems from the fact that in the two media, the two evanescent waves decay at different rates with the distance from the interface. Thus, whatever the relationship between the two evanescent waves at a point in the boundary, it changes by moving up and down in the boundary. It is true that we could envision a coefficient, not unlike the Fresnel transmission amplitude coefficient that sets the connection between the amplitudes of the evanescent waves on two sides of the boundary. Such a coefficient would depend not only on the refractive indices of the two media, but it would also have to change with the distance z from the interface in clear disagreement with the boundary conditions (Eqs. 4.6 and 4.9) connecting the fields across the interface that do not contain such dependence.

The crucial question about evanescent wave that we must address here is whether the evanescent wave actually propagates along the interface. That may appear as an unjustified question. After all, the presence of the factor for propagation along the x-axis in the expressions (Eq. 19.5) seems to provide the definitive answer "yes." However, we already sense multiple problems with accepting the evanescent wave propagation along the interface. For instance, we could anticipate that the propagation of the evanescent wave $E_{12}^{v}(\boldsymbol{x},t)$, as the wave approaches the boundary, would give rise to a reflected evanescent wave moving in the opposite direction:

$$E_R^v(\boldsymbol{x},t) = E_R^v e^{-i\omega t} e^{-2\pi i n_1 k x \sin\theta} e^{-2\pi k z\sqrt{n_1^2 \sin^2\theta - n_2^2}}, \tag{19.6}$$

as signified by the negative sign in the exponent of the propagation factor for the x-axis. That would then give rise to light reflected back from the interface and moving back to the source. Following the hypothesis that evanescent wave actually would reflect at the boundary, we would have the following situation. For the s-polarized beam, the electric fields of all three waves would be parallel both to the interface and to the boundary. How do we determine the electric field of this hypothetical reflected wave E_R^v? It seems that we could simply apply Fresnel equations for normal incidence because the wave is traveling parallel to the interface and perpendicular to the boundary. Since the amplitude reflection coefficient for normal incidence is given by Equation 4.17a, we would have

$$E_R^v = r_{12}^v E_0^v. \tag{19.7}$$

The expression (Eq. 19.7) must be true for both polarizations. Now we run into a problem since for the p-polarized evanescent wave, the electric field vector has a nonvanishing component in the direction of propagation. A regular propagating wave, incident normally to the interface, does not have a component of electric field in the direction of propagation. This therefore casts doubt that the expression (Eq. 19.7) is valid for the p-polarized evanescent wave and, since the expression is the same for the s-polarized wave, it also becomes doubtful that it is applicable for the s-polarized wave.

Another problem with accepting the line of reasoning exemplified by the expression (Eq. 19.7) comes from considering the two evanescent waves, $E_{12}^{v}(\boldsymbol{x},t)$ and $E_{13}^{v}(\boldsymbol{x},t)$. The two fields are given by the expression in Equation 19.5 in terms of the incident wave and the refractive indices of the relevant media. But, the spirit of the expression (Eq. 19.7) would imply that the two fields are related by

$$E_{13}^{v} = t_{23}^{v} E_{12}^{v}. \tag{19.8}$$

The two requirements (Eqs. 19.7 and 19.8) are generally incompatible.

The lesson is that the evanescent wave is not and cannot be treated as a propagating wave. A propagating wave decouples from its source and keeps propagating regardless of what subsequently happens to its source. For instance, if we point a laser to a night sky, turn it on, keep it on for a microsecond, and then turn it off, we have created a 300 meter long beam of light that keeps moving at the speed of light into outer space regardless of what we subsequently do with the laser. This is what we mean when we say that a beam is independent of its source. However, if we turn off the source of the incoming light undergoing supercritical internal reflection, the evanescent wave disappears. There is no evanescent wave outside the area illuminated by the incoming beam. The evanescent wave does not simply forge forward as it arrives to the edge of the illuminated area. Instead, the evanescent wave simply disappears. The evanescent wave needs to be induced and then sustained by the incoming wave. Therefore, it is not clear that the evanescent wave can reflect from the boundary as we have proposed in Equation 19.7 or transmit through as we proposed in Equation 19.8. The evanescent wave is apparently not like a regular wave.

Still, something happens to the evanescent wave as it reaches the edge of the illuminated area. There should exist a transitional zone between the evanescent wave in the illuminated area and no wave outside the illuminated area. This, of course, is not accounted for by our model since we have assumed plane wave solutions and these solutions extend throughout the entire space. Similarly, there should exist a narrow transitional area, centered on the boundary between the two media shown in Figure 19.5, within which the two solutions (Eq. 19.5) smoothly morph into one another.

A propagating wave depends for its propagation on self-regeneration. An oscillating electric field produces an oscillating magnetic field, which in turn produces an oscillating electric field, and so on. We certainly have an oscillating electric field in the evanescent wave behind a totally reflecting interface. So why does it stop at the edge of illuminated area? To better understand this, we need to look yet deeper into the nature of the evanescent wave. It may have seemed that we have investigated the nature of the evanescent wave from all possible angles and that there is really nothing more that we could learn about it, but we see now that what we thought that we understood well about evanescent waves is not providing us with an adequate explanation of what is going on in the above situation and is bringing into question the entire picture that we developed about evanescent waves. A simple question, whether it is propagating along the interface or not, is not satisfactorily answered. We obviously have to look deeper.

First, let us consider the electromagnetic fields of the evanescent wave. Recall that, for our choice of geometry, the totally reflecting interface is in the x-y plane and the evanescent wave is "propagating" in the x-direction. The plane of incidence is the x-z plane and the wave vectors of the incident, the reflected, and the transmitted (evanescent) wave are in the x-z plane. The components of the wave vector of the transmitted wave are

$$\begin{aligned} k_x &= n_2 k \sin\varphi = n_1 k \sin\theta \\ k_z &= n_2 k \cos\varphi = k\sqrt{n_2^2 - n_1^2 \sin^2\theta}. \end{aligned} \tag{19.9}$$

For the p-polarized incident wave, the components of the electric field of the evanescent wave are given by Equation 6.4. We have concluded earlier that the p-polarized evanescent wave is not a transverse wave because it propagates along the x-direction and also has a component of its electric field in the x-direction. This property of the evanescent wave apparently deviates from the general conclusion that we reached when studying Maxwell equations where we concluded from Equation 1.7 that an electromagnetic wave is a transverse wave, that is, that the electric and magnetic fields are mutually perpendicular and both perpendicular to the wave vector that points in the direction of propagation.

Let us, however, notice that, although we said that the propagation is in the x-direction, the wave vector has both x- and z-components. True, for supercritical reflection, the z-component of k is imaginary, but it is not zero.

This observation seems to contradict the statement that the wave propagates in the x-direction, and it implies that something is going on in the z-direction as well. Could it be that the electric field and the wave vector are nevertheless mutually perpendicular, regardless of the imaginary nature of some of their components?

The best way to see if the two vectors are perpendicular is to calculate their scalar product. If the scalar product is zero, the vectors are perpendicular. It is easy to see using Equations 19.9 and 6.4 that the scalar product of $\boldsymbol{E}$ and $\boldsymbol{k}$ is indeed zero. This is yet another surprising result percolating out of the electromagnetic theory. Thus, we can no longer categorically claim that the evanescent field is not a transverse wave.

Since the magnetic field of the p-polarized incident wave for our choice of geometry is perpendicular to the plane of incidence, it is parallel to the y-axis. It is easy to see using Equation 1.7 that the y-component of the magnetic field at the interface is

$$B_{0y} = E_0 \frac{2n_1 n_2^2 \cos\theta}{n_2^2 \cos\theta + n_1\sqrt{n_2^2 - n_1^2 \sin^2\theta}}, \qquad (19.10)$$

while the x- and z-components are zero.

There is another issue that we have to tackle before proceeding with the analysis. We have to clarify how to extract the physical content out of the complex values of the various quantities that we are using. This issue is fairly straightforward when we deal with expressions linear in electromagnetic fields. Then we assume that all that has to be done is to retain the real part of the complex value associated with the field. For instance, in the case of an evanescent wave such as in Equation 5.3,

$$E_t(\boldsymbol{x}, t) = E_{0t} e^{-i\omega t} e^{2\pi i n_1 k x \sin\theta} e^{-2\pi k z\sqrt{n_1^2 \sin^2\theta - n_2^2}},$$

the complex magnitude of the electric field of the evanescent wave can be rewritten as

$$E_t(\boldsymbol{x}, t) = t_{12} E_0 e^{2\pi i n_1 k x \sin\theta - i\omega t} e^{-2\pi k z\sqrt{n_1^2 \sin^2\theta - n_2^2}}, \qquad (19.11)$$

where t_{12} is the Fresnel transmission amplitude coefficient, which is a complex number that can be written in polar form as

$$t_{12} = |t_{12}| e^{i\psi},$$

and E_0 is the magnitude of the incident electric field vector. Therefore, Equation 19.11 could be written as

$$E_t(\boldsymbol{x},t)=|t_{12}|\,E_0e^{2\pi i n_1 kx\sin\theta-i\omega t+i\psi}e^{-2\pi kz\sqrt{n_1^2\sin^2\theta-n_2^2}}. \tag{19.12}$$

The real part of the right-hand side of Equation 19.12 is then simply

$$E_t(\boldsymbol{x},t)=|t_{12}|\,E_0e^{-2\pi kz\sqrt{n_1^2\sin^2\theta-n_2^2}}\cos(2\pi n_1 kx\sin\theta-\omega t+\psi). \tag{19.13}$$

Thus, the complex number notation that we use to describe electromagnetic fields can be easily reduced to a form where we can readily extract the real part for comparison with measured quantities. Note that the above example also included the case where the wave vector had an imaginary component along the z-direction, which caused the propagation factor for the z-direction to metamorphose from oscillatory propagation to exponential decay along the z-axis.

However, the energy density and the Poynting vector contain electromagnetic fields in a bilinear form. Taking the real part of the product of two complex numbers yields more real terms than just the product of the real parts of these numbers. Therefore, with expressions such as energy density and the Poynting vector, we have to extend the rule that we use with fields themselves.

A good rule would be that we require that we use only the real parts of the fields in the expressions for energy density and the Poynting vector. Then, we get a product of two real numbers, which itself is a real number, as is needed.

This said, we can turn back to evaluating the Poynting vector for the evanescent wave. This is similar to what we did back in the expression in Equation 4.14b to calculate the power carried by the transmitted wave. To get the Poynting vector for the case when the transmitted wave is evanescent, we just have to drop the area of the beam cross section from Equation 4.14b to arrive at

$$\boldsymbol{P}_t=\frac{E_t^2}{4\pi}cn_2\frac{2\pi cn_2}{\omega}\boldsymbol{k}. \tag{19.14}$$

By recalling the expressions in Equation 19.9, we immediately see the trouble with the expression (Eq. 19.14). What is the meaning of the z-component of the wave vector which becomes imaginary for evanescent waves? The energy flow is a measurable quantity and thus must be expressible as a real number. An imaginary number cannot be a result of a measurement.

Previously, we were able to select the real part of a complex number to represent a measurable physical quantity. However, Equations 19.14 and 19.9 imply that the flow perpendicular to the interface is purely imaginary. Using our rule, we would have to declare that the energy flow is zero. But it does seem somewhat artificial to simply declare the energy flow to be zero just because the calculated value came out imaginary.

After all, there are electric and magnetic fields at the interface. So let us step back and start from beginning.

The expression for Poynting vector (Eq. 3.3) was derived for fields in vacuum. As we did for the expression in Equation 4.14b, we have to modify it for propagation in a medium. We get

$$\boldsymbol{P} = \frac{cn_2}{4\pi}(\mathbf{E} \times \mathbf{B}). \tag{19.15}$$

For the p-polarized wave that we are considering, the electric field has only x- and z-components and the magnetic field only the y-component. Thus, $\boldsymbol{P}$ has only x- and z-components:

$$\boldsymbol{P} = \frac{cn_2}{4\pi}(-E_z B_y, 0, E_x B_y). \tag{19.16}$$

This is, so far, in agreement with Equation 19.14 since the wave vector $\boldsymbol{k}$ had only x- and z-components. Moreover, B_y is a common factor. So, the difference in nature between the x- and z-components of the Poynting vector for the evanescent field boils down to the difference in nature between the x- and z-components of the electric field. This difference, for a p-polarized evanescent wave in a nonabsorbing medium, is, as we have seen before, that the z-component of the electric field is real, while the x-component is imaginary. Since $i = e^{i\frac{\pi}{2}}$, the imaginary unit simply adds $\pi/2$ to the phase of the complex exponential propagation factor of the x-component of the electric field. So, the real part of the complex propagation factor for E_x switches from cosine to sine. As we found before, the two components oscillate 90° out of phase; E_z oscillates as cosine while E_x oscillates as sine.

We see from Equation 19.16 that the z-component of the Poynting vector contains a cosine term (B_y) multiplied by the sine term (E_x), while the x-component contains the product of two cosine terms. Therefore, P_x is always positive ($\cos^2$), while P_z oscillates between positive and negative values (sin·cos). On average P_z is zero. Energy flows up and down through the totally reflecting interface, but, on average, there is no net flow of energy through the interface. P_x, on the other hand, is always positive.

Note that it helped to step back and reconsider the Poynting vector directly in terms of the fields. The wave vector no longer figures in the expression for the energy flow, and we were able to readily extract the real part of the two components of the Poynting vector. The imaginary unit was absorbed into the phase of the z-component of the electric field. As we have seen, the role of the complex number formalism that we have utilized to describe electromagnetic waves is to simplify formal manipulations and to keep track of various phase shifts.

At first sight, the above finding confirms our earlier conclusion that energy is flowing parallel to the interface. However, as we look at P_x more closely, it appears that the energy flow along the interface is not really a flow due to a propagating wave.

The energy flow seems to be more like a wave on water. Molecules of water only move up and down, but their movement is synchronized in such a way that they produce a wave that travels on the surface of the water. The crest of the wave (and any other wave height) moves with the same speed along the surface. Energy flows only up and down through the interface, but the flow at various points in the interface is phase coordinated to produce a moving wave just like a wave on water. The situation is depicted in Figure 19.6.

This is an important observation and, in many respects, it changes all that we thought we knew about evanescent waves at totally reflecting interfaces. If energy does not flow along the interface, but what flows along the interface instead is the phase of the oscillations of energy flow perpendicular to the interface, then the issue of what happens to the evanescent wave at the edge of the illuminated area of the interface becomes moot. In the points of interface that are not illuminated by the incident beam, there is no energy flow through the interface and, consequently, there is no phase of these in and out oscillations moving along the interface.

The phase moving along the interface must move in synchrony with the point on the interface in which the wave front of the incident wave reaches the interface. We have already concluded that the speed of the evanescent wave along the interface is given by Equation 5.5. A curious fact is that this speed only depends on the refractive index of the

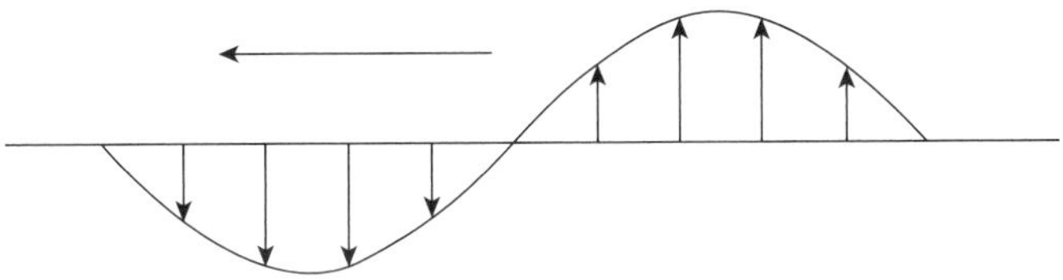

Figure 19.6 The energy flow through the interface depicted by perpendicular arrows is phase coordinated to produce an energy wave that moves along the interface (horizontal arrow).

incident medium (and not the medium in which the evanescent wave actually propagates) and on the angle of incidence of the incoming wave. How could that be possible? If the evanescent wave was a real traveling wave in the medium beyond the interface, its speed of propagation would have been dependent on the refractive index of that medium and it would not have anything to do with the angle of incidence of the incoming wave. Indeed, the result that the speed of propagation of the evanescent wave depends on the angle of incidence of the incoming wave should have sounded the alarm bell right then and there. Maxwell equations require that electromagnetic waves propagate through a medium with a very specific speed determined entirely by the refractive index of the medium. They do not leave any room for other speeds of propagation.

We now see that there is nothing physically moving along the interface in the first place, so there is nothing that would have to continue to move beyond the edge of the illuminated area. This finally solves the original puzzle of why the evanescent wave, traveling along the interface through an optically inhomogeneous material such as powder, is not scattered at the various grain boundaries in the powder, as it should be if it was a true propagating wave.

Another analogy to the evanescent wave propagation that could solidify this new image of the evanescent wave is that of a threaded rod rotating slowly around its axis but otherwise stationary. If we look at the thread on the screw, it apparently moves along the axis. If the rod changes the sense of its rotation, the thread moves the opposite way. However, we already said that the screw is really not moving along the axis at all. Each material point of the screw just rotates around the axis in a plane perpendicular to the axis. The "motion" of the thread is just an illusion. At the beginning and at the end of the rod, the thread just stops moving and the apparent motion of the thread simply ends.

This is a fascinating revision of our understanding of evanescent wave propagation. We entered this chapter believing that energy does not flow through the interface and that the evanescent wave, and thus the energy associated with it, propagates parallel to the interface. Now we are turning that upside down and are concluding that energy is really not flowing parallel to the interface, but that there is an oscillatory flow of energy back and forth through the interface. All that is moving along the interface is the phase of this oscillatory flow. In the end, the motion of the phase of the oscillatory flow of energy in some sense represents the flow of energy, just as a wave on the surface of water carries energy, although the motion of molecules of water is only up and down, that is, perpendicular to the energy flow.

20 Energy Flow at a Totally Internally Reflecting Interface

20.1 ENERGY CONSERVATION AT A TOTALLY REFLECTING INTERFACE

Let us reexamine the phenomenon of total internal reflection from the point of view of conservation of energy. We concluded that, since the reflectance for total internal reflection in the case of nonabsorbing media is 100% (Eq. 4.23), all the incident energy is reflected from the interface and thus energy is conserved. This, of course is true, but the details are more subtle than this statement would imply.

We commented that the reflectance amplitude coefficient becomes a complex number of unit magnitude making the reflected wave oscillate at a constant phase offset ψ with respect to the incoming wave. We let the statement stay without further examining it, but if the two fields are not oscillating in phase, the incoming energy density flow,

$$P_{in}(\boldsymbol{x},t)=\frac{|\boldsymbol{E}_0|^2}{4\pi}\cos^2(2\pi n_1\boldsymbol{kr}-\omega t), \tag{20.1}$$

and the totally internally reflected ($|r_{12}|^2=1$) outgoing energy density flow,

$$P_r(\boldsymbol{x},t)=\frac{|\boldsymbol{E}_0|^2}{4\pi}\cos^2(2\pi n_1\boldsymbol{kr}+\psi-\omega t), \tag{20.2}$$

do not match in phase! If we set the origin of the coordinate system in the interface, the energy flow balance in the origin ($\boldsymbol{r}=0$) is simply

$$\begin{aligned}P_r(\boldsymbol{0},t)-P_{in}(\boldsymbol{0},t)&=\frac{|\boldsymbol{E}_0|^2}{4\pi}\left(\cos^2(\psi-\omega t)-\cos^2(\omega t)\right)\\&=\frac{|\boldsymbol{E}_0|^2}{4\pi}\sin(\psi)\sin(2\omega t-\psi).\end{aligned} \tag{20.3}$$

Internal Reflection and ATR Spectroscopy, First Edition. Milan Milosevic.

Thus, there are intervals of time when the reflected energy flow is smaller than the incident energy flow and there are intervals of time when the reflected energy flow exceeds the incident energy flow. While, on average, the energy flow (Eq. 20.3) balances out (the average value of the sine is zero), the question is what happens with the momentary energy flow that is not balanced? Where does the energy go when the reflected energy flow is smaller than the incident energy flow, and where does it come from when the reflected energy flow exceeds the incoming flow?

Energy must be conserved at every moment, not only on average. Thus, we expect that the excess energy flows through the interface, taking the excess away when the reflected energy flow is smaller than the incoming flow and supplying the excess energy when it is larger. On the other side of the interface is the evanescent wave. Thus, the energy flow imbalance is pumped into and out of the evanescent wave. The evanescent wave serves as a temporary storage of electromagnetic energy that takes in the excess of the energy flow and returns it out, depending on the phase shift between the totally reflected and the incoming wave. We now see that, in order for a totally reflected electromagnetic wave to have a phase shift with respect to the incoming wave, there must exist a temporary storage of electromagnetic energy just behind the totally reflecting interface. In other words, the evanescent wave acts as an energy flux capacitor.

The situation is depicted in Figure 20.1. The incident (Eq. 20.1) and reflected (Eq. 20.2) energy flows enter through the surfaces, shown as a dashed line. The flow through the interface is given by the z-component of the expression (Eq. 19.16). Note that the surfaces through which the incident and reflected light flow have the same areas and that the area

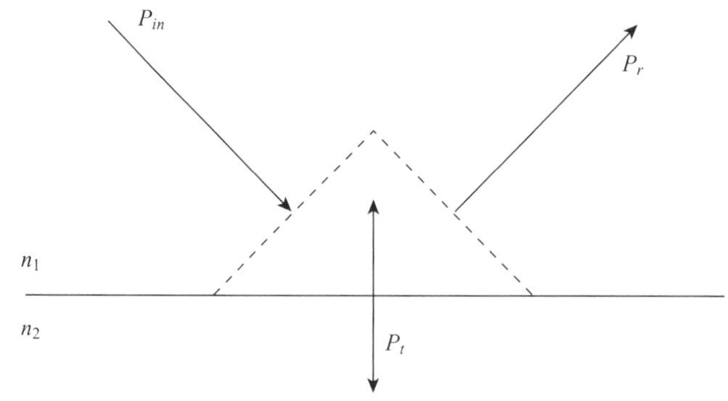

Figure 20.1 Incoming energy flow must equal reflected flow plus the flow through the interface. The flow through the interface oscillates between outward and inward.

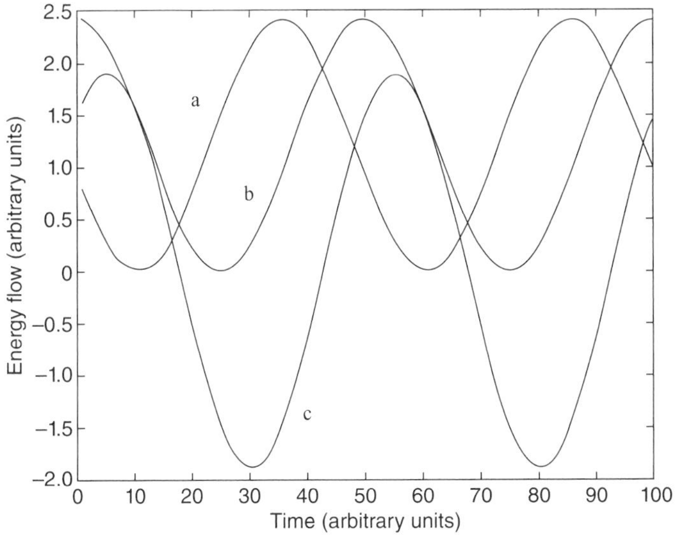

Figure 20.2 Energy flow at a totally reflecting interface. Incoming energy flow (a), reflected energy flow (b), and the flow through the interface (c).

of the flow through the interface is larger than the incident flow by a factor of $1/\cos\theta$. As we have seen, a phase shift between the incoming and reflected waves causes an instantaneous imbalance in the energy flow, and we want to show that the imbalance is cured by the flow through the interface.

On the face of it, it does seem unlikely that the imbalance (Eq. 20.3) is cured by the z-component of Equation 19.16. The expression for Equation 19.16 includes a transmission amplitude coefficient, which is a complex number for supercritical internal reflection, and it seems highly unlikely that all the phase shifts and magnitudes conspire to balance the energy flow at the interface.

Figure 20.2 shows the time dependence of the incoming (a), the reflected (b), and through-the-interface flow (c). We see that the reflected energy flow is the same, on average, as the incoming, except that it is phase shifted in time. Both the incoming and reflected energy flows are always positive. The through-the-interface flow is both positive and negative. Amazingly, it exactly balances the difference between the reflected and incoming energy flows.

It should not be such a surprise that the theory obeys energy conservation in this special case since it obeys it on the most general level. It is just that the expressions did not seem likely to cooperate with energy conservation. This is easily demonstrated to be true for both polarizations and for any supercritical angle of incidence. It also finally confirms the picture of the evanescent wave that we put forward in the discussion surrounding Figure 19.6. It also confirms that energy flows along the interface, although this flow is not due to an electromagnetic wave that travels along the interface. All that travels along the interface is the phase of the coordinated bobbing of energy back and forth through the interface. This does superficially imply the transport of energy along the interface, but that transport is not facilitated by a traveling electromagnetic wave. Since there is no traveling wave, there is nothing to scatter in the powdered medium behind the interface.

This finally resolves our initial question of why the evanescent wave does not scatter in a powered medium.

20.2 SPEED OF PROPAGATION AND THE FORMATION OF AN EVANESCENT WAVE

Let us observe one more intriguing feature regarding the speed of propagation of the evanescent wave. Maxwell equations require that the speed of light in the medium of refractive index n_2 is always $c_2 = c/n_2$. However, as we have already seen, the speed of the evanescent wave is $c_e = c/(n_1 \sin\theta)$. At the critical angle, $\sin\theta = n_2/n_1$ and the speed of the evanescent wave is equal to the speed of light in medium 2. This is an intriguing result. It implies that the evanescent wave exactly at the critical angle actually becomes a true propagating wave. In other words, it can decouple and continue to propagate into the unilluminated portions of the interface. For angles of incidence larger than the critical, the speed of the evanescent wave is smaller than the speed of light in medium 2. Incidentally, this offers yet another way to define the critical angle and another way to visualize the distinction between the phenomena of subcritical and supercritical internal reflections as illustrated below. Figure 20.3 illustrates how a transmitted wave is formed where the wavelets, induced by an incoming wave at the interface, form a coherent wave front. In all other directions, the waves induced by the incoming wave at the interface do not superimpose coherently. They mutually interfere, each wavelength with its own phase, and equal numbers of positive and negative contributions add up to zero average intensity.

With supercritical incidence, as shown in Figure 20.3b, coherent superposition cannot occur and waves add incoherently everywhere in medium 2 except very near the interface itself. Near the interface, the field is dominated by the wavelets induced in the adjacent areas on the interface, which, because they are very close, are mutually coherent.

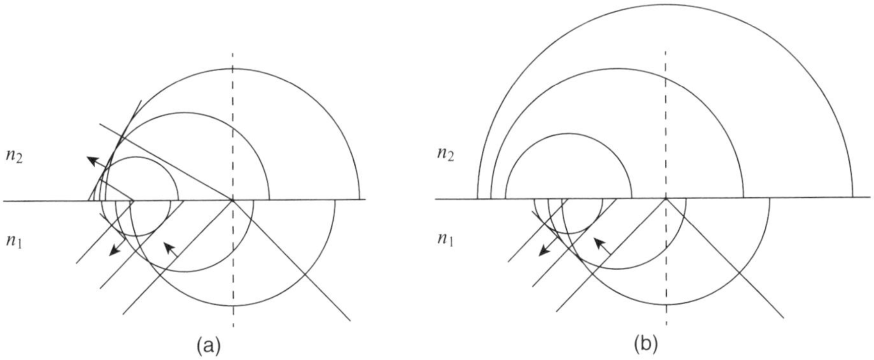

Figure 20.3 A point in which a wave front of the incoming wave hits the interface travels along the interface at speed $c/n_1 \sin\theta$. According to the Huygens principle, a propagating wave induces a new wave in every point in space reached by the propagating wave. These new waves oscillate with the same frequency and are in phase with the propagating wave. The successive waves originating at points in the interface propagate at speeds $c_1 = c/n_1$ in medium 1 and $c_2 = c/n_2$ in medium 2. In medium 1, the waves induced in the interface by the incoming wave form a common tangential wave front of the reflected wave. The situation in (a) corresponds to the case where speed $c/n_1 \sin\theta$ is larger than c_2. In that case, the waves originating in interface and propagating in medium 2 form a common tangential wave front of the transmitted wave. The situation in (b) correspond to the case where the speed of light in medium 2 is larger than $c/n_1 \sin\theta$ (i.e., the supercritical incidence) and waves induced in the interface by the incoming wave never catch up one with another. No common wave front can be formed in medium 2, and hence, there is no propagating wave in medium 2.

The yardstick that defines near and far is, as always, a wavelength of light, and near means less than and far means more than the wavelength of light. This description shows vividly how an evanescent wave is formed and why it has a significant amplitude only very near the interface.

21 Orientation Studies and ATR Spectroscopy

21.1 ORIENTED FRACTION AND DICHROIC RATIO

Suppose there is an angle θ between a molecular bond orientation and an applied field. The orientation of the bond is kept fixed by the rigidity of the material. We also assume that a dipole moment can be induced only in the direction of the bond. Thus, it is the projection of the electric field $\boldsymbol{E}\cos\theta$ onto the bond direction that induces the dipole moment. Therefore, in Equations 2.6 and 2.8, instead of $\boldsymbol{E}$, we should have $\boldsymbol{E}\cos\theta$. To see how the absorption coefficient depends on the angle θ, we can calculate the energy loss per unit time (i.e., the absorbed power) due to the frictional force $F = -m\gamma v$ in Equation 2.6 as

$$P_a = \frac{1}{T}\int_0^T F dx = \frac{1}{T}\int_0^T F v dt = \frac{m\gamma\omega^2 X_0^2}{2} = \omega^2\gamma \frac{\frac{q^2}{2m} E^2\cos^2\theta}{\left(\omega_0^2 - \omega^2 + i\gamma\omega\right)^2}. \quad (21.1)$$

The important part of the above expression is the appearance of the square of $\cos\theta$. This implies that the absorbance of the bond is maximal when the electric field of the electromagnetic wave is aligned with the bond direction ($\cos 0 = 1$) and is zero when the fields is perpendicular to the bond direction ($\cos 90° = 0$). Equation 21.1 shows the power absorbed by a single bond of a single molecule. To find the total power absorbed per unit volume, the above expression has to be summed over all N molecules in unit volume and over all the bonds in the molecule. However, in summing up over all the molecules, we cannot assume that all the molecules are oriented in space in the same way. Thus, we have to find the average value of the cosine squared:

$$s\left\langle\cos^2\theta\right\rangle = \frac{1}{N}\sum_{i=1}^{N}\cos^2\theta_i.$$

Internal Reflection and ATR Spectroscopy, First Edition. Milan Milosevic.

Let us first consider two extreme cases. First, let us assume that all the molecules are indeed oriented identically. Then the average angle is the same as the angle for every molecule and the absorbance of such a sample is at maximum when the electric field is oriented in the same direction as the bond, and zero when it is perpendicular. On the other hand, if the molecules are randomly oriented, not preferring any particular direction, the result of averaging must be the same for every direction. Since there are three directions (x, y, z) and $\cos^2\theta_x + \cos^2\theta_y + \cos^2\theta_z = 1$ for any individual molecule, it follows that, for randomly oriented molecules, the average value is

$$\langle \cos^2 \theta \rangle = 1/3. \tag{21.2}$$

It can be shown that, for the purpose of analysis of molecular orientations by absorption spectroscopy, any distribution of molecules can be modeled by a linear combination of the above two extreme cases. In a way, this clearly shows the limits on the richness of information that orientation studies can provide.

The degree of orientation can be quantified by considering that a certain fraction f of molecular bonds is perfectly oriented in a particular direction, and the remainder, $(1 - f)$, is oriented randomly. The absorbance A_1 is measured for the field oriented along the direction of orientation and the absorbance A_2 is measured for the field perpendicular to that direction. Therefore, for the first measurement, the average of cosine squared is one and for the second measurement, it is 1/3.

The absorbance A_1 is due to the fraction f of the molecules oriented along this direction plus $(1 - f)/3$ of the randomly oriented bonds. The absorbance A_2 is only due to $(1 - f)/3$ of the randomly oriented bonds, and thus the ratio of the two is

$$D = \frac{A_1}{A_2} = (1 + 2f)/(1 - f). \tag{21.3}$$

This allows the calculation of the fraction f as

$$f = \frac{D - 1}{D + 2}. \tag{21.4}$$

This method could be extended to allow the measurement of absorption along the z-direction as well. By tilting the sample so that the angle of incidence is no longer normal, the electric field is then no longer perpendicular to the z-direction (thickness) but acquires a component along that direction. This component is a function of the angle of incidence and also of the refractive index of the sample. The absorbance measured in this situation will be due to the electric field component along the z-direction and the electric field component in the x-y plane. We can rotate the sample so the x-component of the electric field is zero. Then, by knowing the absorbance in the y-direction from the previous measurement, we can subtract it from the measured absorbance to arrive at the absorbance for the z-direction. Analogously, we can do the measurement of the sample absorbance for that rotation of the sample that makes the y-component of the electric field inside the sample vanish.

Imagine a sample made in the shape of a long thin film, like a tape. We denote the direction along the thickness as the z-axis, the width of the tape as the y-axis, and the length as the x-axis. We expect that the tape's x-direction is the so-called machining direction. If the manufacturing process was not symmetric with respect to the x- and y-directions, then the molecular structure of the film may reflect that asymmetry. We want to compare the strengths of absorption peaks for the cases when the electric field of incident light is polarized along the x-direction and along the y-direction. A way to do it by using transmission spectroscopy is simple. We polarize the incident beam so that it is in the vertical plane. The electric field then oscillates in the up-down direction. We place the sample in the beam and orient it so that the y-direction of the sample is up-down and collect the spectrum A_y. Then, we rotate the sample so that the x-direction is up-down and collect the spectrum A_x. If the two spectra are identical, it would be reasonable to conclude that the sample is isotropic, that is, equal in all directions. However, if we find that a particular absorption peak centered at a wave number, k_1, is much stronger in A_x than in A_y, the conclusion would be that the sample is anisotropic and that the particular chemical bond whose absorption is centered at k_1 is preferentially oriented along the x-direction. The dichroic ratio A_x/A_y can thus be measured for every absorption band of the material. We may find that certain bonds are preferentially oriented along the x-direction and some bonds along the y-direction. Clearly, this reflects the underlying molecular structure and can be used to elucidate the molecular structure of the unknown molecules. Or, a certain degree of orientation may be desired to give the film some desirable mechanical properties. Thus, the measurement of the dichroic ratio of the film can be used as a quality control method in film production.

21.2 ORIENTATION AND FIELD STRENGTHS IN ATTENUATED TOTAL REFLECTION (ATR)

Transmission spectroscopy, and the experimental method described above, is limited to optically thin samples. Attenuated total reflection, however, is not restricted by sample thickness. In ATR, the absorption results from the interaction of the sample and the evanescent wave. The electric fields of the evanescent wave at the interface are given by Equations 6.2 and 6.4. However, since here we are not interested in the phase shift between the incident field and the evanescent wave, we can use the absolute value of those expressions:

$$E_y = \frac{2n_1 \cos\theta}{\sqrt{n_1^2 - n_2^2}} E_0^s \tag{21.5}$$

and

$$\begin{aligned} E_x &= \frac{2n_1 \cos\theta \sqrt{n_1^2 \sin^2\theta - n_2^2}}{\sqrt{n_1^2 - n_2^2}\sqrt{(n_1^2 + n_2^2)\sin^2\theta - n_2^2}} E_0^p \\ E_z &= \frac{2n_1^2 \sin\theta \cos\theta}{\sqrt{n_1^2 - n_2^2}\sqrt{(n_1^2 + n_2^2)\sin^2\theta - n_2^2}} E_0^p. \end{aligned} \tag{21.6}$$

For the s-polarized incident light, only E_y exists, and both E_x and E_z are zero. For the p-polarized incident light, E_y is zero, while E_x and E_z are nonzero.

Let us consider how the previous gedanken experiment with the thin film that we performed above using transmission spectroscopy could be accomplished with ATR. To analyze the sample in the x-y direction, we could use s-polarized incident light, then place the sample in contact with the ATR element; first so that its x-direction is along E_y, measure the absorbance A_x, and then rotate the sample so that its y-direction is along E_y and then measure A_y. Since the same field (E_y) is used for both measurements, the ratio of the measured absorbances is just the ratio of the respective average cosines squared:

$$\frac{A_x}{A_y} = \frac{\langle\cos^2\theta_x\rangle |E_y|^2}{\langle\cos^2\theta_y\rangle |E_y|^2} = \frac{\langle\cos^2\theta_x\rangle}{\langle\cos^2\theta_y\rangle}. \tag{21.7}$$

The absorbance depends both on the field strength for a particular direction and on the average cosine squared between the bond and the electric field direction. In order to determine the geometric information on the molecular structure of the sample, the ratio of the field strengths squared for the two directions must be extracted from the ratio of the measured absorbances to arrive at the ratio of cosines squared for the two directions. With the s-polarized incident beam, we only had the electric field oscillating in one direction, and we oriented the sample to bring the two directions in the sample, one at a time, in alignment with the field. So the electric field strength was the same for the two directions and it simply cancelled out in the ratio. In a sense, this experiment was quite similar to the transmission experiment described earlier. However, if we use the p-polarized beam instead, we can get information on all three directions, but the field strengths for the two directions are not the same. If the sample is oriented so that the x-direction in the sample coincides with E_x, the absorbance measured would be

$$A_1 = C(\langle\cos^2\theta_x\rangle|E_x|^2 + \langle\cos^2\theta_z\rangle|E_z|^2).$$

If the sample is oriented so that the y-direction in the sample is along E_x, the absorbance measured would be

$$A_2 = C(\langle\cos^2\theta_y\rangle|E_y|^2 + \langle\cos^2\theta_z\rangle|E_z|^2),$$

where C is a constant. Thus, the ratio of A_1 and A_2 becomes

$$\frac{A_1}{A_2} = \frac{\langle\cos^2\theta_z\rangle + \langle\cos^2\theta_x\rangle\dfrac{|E_x|^2}{|E_z|^2}}{\langle\cos^2\theta_z\rangle + \langle\cos^2\theta_y\rangle\dfrac{|E_y|^2}{|E_z|^2}}. \tag{21.8}$$

The field ratios are easily calculated from Equations 21.5 and 21.6. The two ratios (Eqs. 21.7 and 21.8), together with the generally valid relationship between the direction cosines,

$$\langle\cos^2\theta_x\rangle + \langle\cos^2\theta_y\rangle + \langle\cos^2\theta_z\rangle = 1,$$

suffice to extract all three cosines squared for a particular sample. This calculation can be done for any wavelength. The result for a different absorption peak would be reflective of the orientation of the particular chemical bond responsible for that absorption peak.

Note that we have assumed that the refractive index of the sample n_2 is the same in all directions. For oriented samples, this is generally not true. The refractive indices of anisotropic samples are generally direction dependent. However, for most samples, this dependence of the refractive index of the sample on the direction is, justifiably or not, generally ignored.

It is, however, important to realize that the information obtained from the dichroic measurements as outlined above is not very rich. For instance, a hypothetical sample that has a particular molecular bond aligned in such a way that a third of the molecules have this bond along the x-axis, a third along the y-axis, and a third along the z-axis would yield identical results as a sample of the same molecules with bonds oriented perfectly randomly in all directions. This illustrates the relative crudeness of the information obtained through orientation studies. Still, the information obtained can be very useful, especially with highly oriented samples for which the vanishing absorbance of a particular bond would clearly indicate that the bond responsible for the absorption is aligned perpendicular to the electric field of the wave propagating in the material. This is a very unambiguous statement about the molecular structure of the sample and its molecular alignment.

22 Applications of ATR Spectroscopy

Attenuated total reflection spectroscopy (ATR) became the dominant spectroscopic technique in the infrared (IR) spectral region for the simple reason that it allows spectral acquisition of all types of samples without virtually any sample preparation. Below is a brief summary of the advantages of ATR spectroscopy for different types of applications and for different types of samples.

22.1 SOLID SAMPLES

It is possible to analyze solid samples using transmission spectroscopy by mixing a small piece of sample with a nonabsorbing matrix such as KBr, grinding it into a fine powder and pressing it into a pellet. A transmission spectrum of a solid sample can be collected, but it takes a lot of skill in the art of sample preparation to make a good pellet. The sample is afterward difficult to recover, and, sometimes, the sample–matrix interaction influences peak shapes and positions.

To collect a spectrum of the same sample by ATR, the sample is pressed against the surface of the ATR element and the spectrum acquired. With single reflection diamond ATR, there is no challenge involved.

22.2 LIQUID SAMPLES

Liquid samples are obviously easy to analyze with ATR spectroscopy since liquid automatically forms full contact with the ATR element. Liquid is also easily analyzed using transmission spectroscopy, but ATR spectra naturally produce short effective path lengths required for

Internal Reflection and ATR Spectroscopy, First Edition. Milan Milosevic.

some types of liquid samples such as aqueous solutions. There are no interference fringes in the spectra. Flow-through ATR cells do not require flowing a sample through a narrow gap—a particular advantage with viscous liquids.

22.3 POWDERS

Powders are a type of sample that has been difficult to analyze using transmission spectroscopy. As with solids discussed above, one can mix a bit of powdered sample into a nonabsorbing matrix such as KBr and press a pellet. But, as we saw, the preparation of pellets is an involved procedure.

Mixing the powdered sample with KBr powder and using diffuse reflection spectroscopy became very popular a few decades ago. The procedure avoids making pellets and that was one of the main reasons for its popularity. Increasingly, however, spectroscopists chose to replace the diffuse reflection technique with simply taking the ATR spectra of powdered samples directly and without any sample preparation. Samples are analyzed neat and are easily recovered after the analysis. Also, the spectral interpretation of ATR spectra of powders is the same as for any other type of samples.

22.4 SURFACE-MODIFIED SOLID SAMPLES

Solid materials that have a very thin surface layer modified in some way are difficult to analyze using transmission spectroscopy. One can slice a thin surface layer from the sample and analyze it by transmission spectroscopy, but that requires quite sophisticated sample preparation techniques and is destructive to the sample.

Attenuated total reflection spectroscopy is a tailor-made technique for this type of sample since it naturally probes the very surface layer of the sample in contact with the ATR element. One can choose the ATR material and select the angle of incidence to further optimize the ATR technique for a particular sample.

22.5 HIGH SAMPLE THROUGHPUT ATR ANALYSIS

Fourier transform infrared spectroscopy is a very sensitive analytical technique. Just propagating through the ambient atmosphere between the source and detector in a spectrometer, the spectrometer's IR beam is absorbed by the atmospheric IR absorbing gases such as water vapor and CO_2. These atmospheric absorptions vary in intensity as the concentrations of the absorbing components vary. Thus, as the ratio of background and sample single beams is calculated to find the spectrum, the unbalanced absorptions of water vapor and CO_2 are superimposed onto the sample spectrum. These interferences may be considerable, particularly for weakly absorbing samples.

The standard way to suppress these atmospheric interferences is to use a nonabsorbing gas such as nitrogen to purge the interior of the spectrometer. However, opening the spectrometer sample compartment to introduce the sample exposes this interior atmosphere within the spectrometer to the outside air. Purging the spectrometer with a flow of nitrogen restores the nonabsorbing atmosphere inside the spectrometer, but this procedure can take between 15 and 30 minutes to complete. This limits the rate at which samples can be analyzed.

In ATR spectroscopy, the interaction of light with the sample occurs through the evanescent wave. The light from the spectrometer is completely confined within the ATR element. This offers the possibility to enclose the entire optical path of the light beam within the spectrometer, leaving only the top surface of the ATR element exposed to the outside. The spectrometer beam thus remains at all times enclosed in the controlled ambient of the spectrometer. A sample can now be placed on the ATR element, the spectrum acquired, the sample removed, the ATR element cleaned, and everything is ready for a new sample. This greatly increases the rate at which samples can be analyzed.

22.6 PROCESS AND REACTION MONITORING

One can imagine an ATR sensor embedded in the wall of an industrial process vessel. A spectrometer interfaced to this ATR element would enable the monitoring of the chemical composition of the liquid in the vessel. Or alternatively, a hollow tube with an ATR sensor embedded at one end could be used to bring light through the tube to the ATR sensor and back. Such a tube, interfaced on the other end to a spectrometer, could be inserted into a process vessel through one of the standard access ports. As the chemical reaction in the vessel proceeds from raw ingredients to the final product, changes in the chemical composition of the process liquid are reflected in the acquired ATR spectra. Detailed reaction dynamics can be monitored, process parameters adjusted to optimize the speed and yield of the reaction, the end point of the reaction can be accurately determined, and so on.

In some cases, alternative reaction pathways, starting from the same initial state of reaction, are possible. Which particular pathway takes place may be determined by subtle variations in the values of the process parameters. Studying reactions in real time enables the selection of process parameters that lead to the reaction pathway that provides the desired outcome in the fastest and most economical way. Reaction monitoring is a powerful tool for process optimization.

APPENDIX A
ATR Correction

We have seen that spectra collected by transmission and attenuated total reflection (ATR) spectroscopy depend on the optical constants of the sample and some experimental parameters. When transformed into absorbance, both spectra show a series of characteristic peaks at approximately the same positions and of similar shapes. The main difference between absorbance transformed transmission (Eq. 1.12),

$$A = 0.434 \cdot 4\pi k \kappa d, \tag{A.1}$$

and absorbance transformed ATR spectra (Eq. 7.4), which, after incorporating the definitions in Equations 8.1 and 8.2, simplify to

$$\begin{aligned} A_S &= 0.434 \cdot 4\pi d_{\text{eff}}^{s} \kappa \\ A_p &= 0.434 \cdot 4\pi d_{\text{eff}}^{p} \kappa, \end{aligned} \tag{A.2}$$

comes from the explicit wave number dependence of the transmission spectra as evidenced by the presence of wave number k in Equation A.1. Thus, to make ATR spectra appear similar to transmission spectra, all one needs to do is to multiply the ATR spectra by the wave number k. This increases the apparent absorbance at the shorter wavelengths and makes the ATR spectra appear transmission-like. In practice, since in the mid-IR the wave number takes large values (from 400 to 5000), the absorbance transformed ATR spectra are multiplied by k/k_0; that is,

$$A^{ATRcorr} = A^{ATR} \frac{k}{k_0}, \tag{A.3}$$

where k_0 is the wave number at which the absorbance transformed ATR spectrum has the same value as the absorbance transformed

Internal Reflection and ATR Spectroscopy, First Edition. Milan Milosevic.

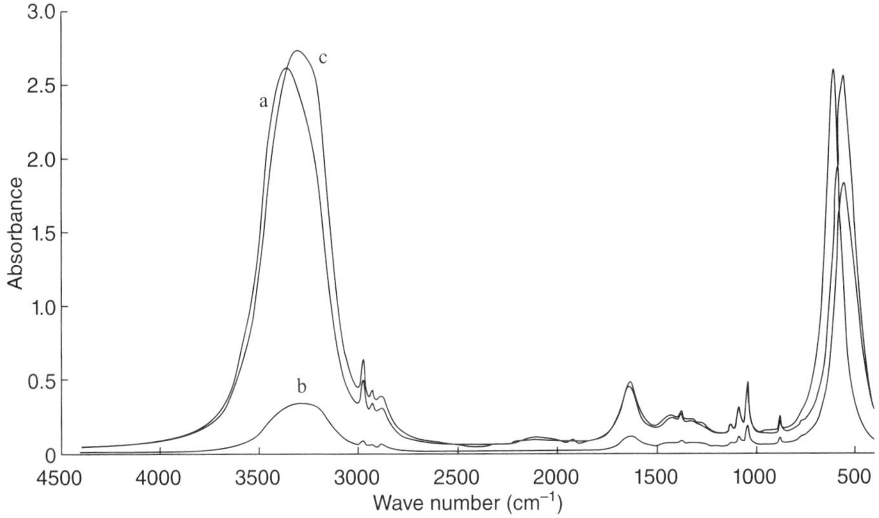

Figure A.1 Comparison of transmission (a) ATR (b) and ATR corrected (c) spectra of a sample.

transmission spectrum. In some implementations of the correction algorithm, the wave number k_0 can be selected by the user, and in some implementations, it has a fixed value (i.e., 1000 cm^{-1}).

As long as the effective thickness of a sample does not change much as a function of wave number, the ATR correction yields ATR spectra that reasonably well resemble transmission spectra. The free parameter k_0 can be used to appropriately scale the corrected ATR spectrum for library search or for comparison to actual transmission spectrum.

Figure A.1 illustrates the effect of the ATR correction. An absorbance transformed transmission spectrum of a sample is compared with the absorbance transformed ATR spectrum of the same sample. As it would be expected, the ATR correction works extremely well for weak absorptions. The agreement between the transmission and the ATR corrected spectra becomes less satisfactory for strong absorption peaks. This is because the changes to the refractive index of a sample near a strong peak are large, and thus the effective thickness of the ATR measurement is no longer constant. We see that the peak strengths are reasonably well corrected, but the peak positions and peak shapes are not corrected at all. There is no doubt that the ATR correction improves the similarity between the transmission and ATR spectra, but we also see that the correction is far from perfect and that for strong peaks, there remain significant differences between transmission and ATR corrected spectra.

The principal benefit of the ATR correction is that it improves automated searching of spectral libraries. Spectral libraries are vast collections of spectra of known materials. They are useful tools for the identification of unknown samples. The spectrum of an unknown sample is compared to a huge number of spectra of known substances. Sophisticated mathematical algorithms are used to evaluate similarities between an unknown spectrum and each library spectrum. A numerical score is assigned to each comparison and the library spectra with the highest scores are displayed for visual comparison. This generally allows quick sample identification. The spectra collected in spectral libraries are generally acquired using transmission spectroscopy. Therefore, the use of ATR corrected ATR spectra improves the similarity scores, making them more meaningful.

APPENDIX B
Quantification in ATR Spectroscopy

We stated that the quantitative analysis of samples is one of the main utilizations of spectroscopy. We discussed Beer's law as the principal building block on which quantitative spectroscopy is built. However, as we have seen, Beer's law is not really a law. It is just an approximation that works well in many cases. Figure B.1 shows the attenuated total reflection (ATR) spectra of a sample whose concentration changes from 0% to 100% in steps of 10%. We are focusing on an isolated peak at $1000\,cm^{-1}$. The spectra in Figure B.1 are, of course, numerically simulated, but the example is realistic and typical of situations encountered in real life. As often encountered, the peak not only changes height but it also shifts position. The position shift is due to the real part of the refractive index of the sample becoming stronger with increasing concentration and thus the effective thickness increasingly deviating from a constant value. One way we could use the information contained in Figure B.1 would be to measure the absorption peak heights associated with different concentrations and to form a graph of concentration versus peak height. The graph would then yield an expression of the type

$$c = f(A), \tag{B.1}$$

where f is some function of measured absorbance. Ideally, f would be linear; that is, we would have

$$f(A) = a + bA, \tag{B.2}$$

where a and b are parameters determined from experimentally measured values. However, we may decide that a linear form is not the best

Internal Reflection and ATR Spectroscopy, First Edition. Milan Milosevic.

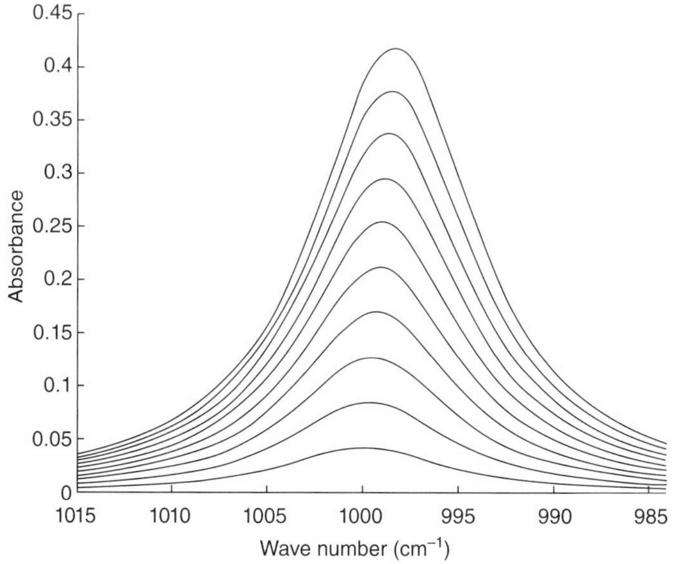

Figure B.1 Concentration versus absorbance dependence derived from the spectra of Figure B.2 (points) and the linear and quadratic fit (almost indistinguishable solid lines) for the experimental values.

choice for the concentration versus absorbance relationship and would like to use a quadratic fit. In that case, the functional form for $c(A)$ is

$$c(A) = a + bA + cA^2. \tag{B.3}$$

Both forms (Eqs. B.2 and B.3) are easy enough to use to calculate the concentration for an unknown sample from the measured absorbance of that sample. If practical considerations are as described, the quantification procedure based on either Equation B.2 or B.3 would be straightforward to implement. The curves for both the linear and the quadratic fit are shown in Figure B.2. The two curves are so close that it is hard to distinguish between them, and both represent excellent fits for the experimental points. So, the absorbance at a particular wave number is read from an experimentally measured sample containing the unknown concentration of the substance responsible for the absorption peak; the value is inserted into the expression (Eq. B.2 or B.3); and the unknown concentration is numerically evaluated.

Note that in the above, we used the height of the absorbance peak to obtain the unknown concentration. One problem that we encounter immediately is that the position of the peak shifts with increasing absorbance, so we have to choose whether we are going to use the absorbance value at a particular wave number (say, 1000 cm^{-1} in our

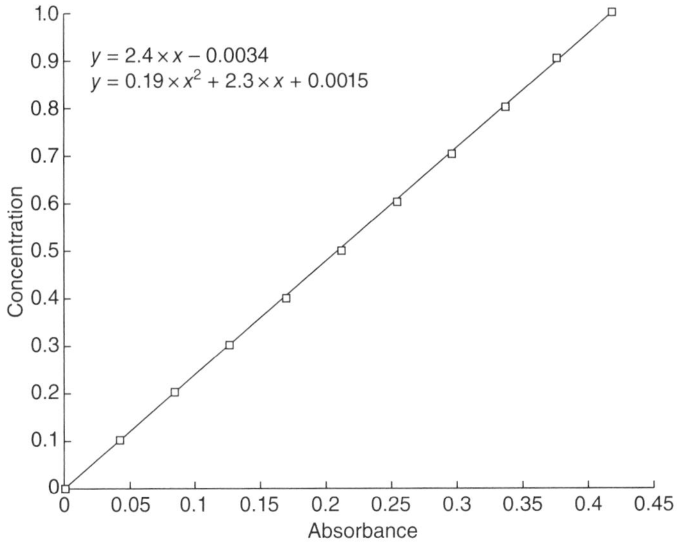

Figure B.2 ATR spectra of a sample whose concentration changes from 0% to 100% in steps of 10%.

example) or the absorbance at the peak maximum. Also, notice that absorbance increases with increasing concentration at any value of the wave number near the peak. So, using the value at the peak maximum seems somewhat arbitrary. For weak absorbance, using the highest absorbance to evaluate the concentration makes sense based on the result (Eq. 1.22) that we found for signal/noise (S/N) of a spectroscopic measurement. However, for absorbance values above 0.4, the S/N curve starts decreasing with increasing absorbance, and there is no reason to continue to stick with the absorbance value at the peak maximum. We could switch to a neighboring value of wave number in which the absorbance is somewhat weaker.

Clearly, concentration information is spread over many points in the spectrum, and restricting the quantification to a single point in the spectrum necessarily leaves out the information carried by these left out points. An alternative approach is not to use the value of absorbance at a single wave number but to use the area under the absorbance peak instead:

$$\int A(k)dk, \tag{B.4}$$

and to follow the above procedure. One benefit of using the area under the peak instead of just a single value of absorbance is that increasing the number of experimental points increases the value of the integral

(Eq. B.4). If the absorbance values for all the wave numbers included in the integral were equal, the integral would be proportional to the number of points N included in the integral. The noise values, however, are random and oscillating between positive and negative values, so the integral of the noise would be proportional to the $\sqrt{N}$. Therefore, the S/N of the measurement that uses the area under the peak increases with the square root of the number of points included in the integral. The benefit of including more and more points in the integral (i.e., widening the range of integration) is, however, in practice limited by the absorbance rapidly decreasing on both sides of the peak maximum. Still however, the area under the peak approach uses much more information present in the measured spectrum than does the absorbance at a single wave number. The question naturally arises if there is a way to use the entire information from a spectrum to quantify the concentration of a component. The answer is yes. This approach to quantification in spectroscopy is described next.

Imagine we are interested in a mixture of two ingredients, A and B. Assuming a linear relationship between the absorbance and the concentration, the spectrum of a mixture is a combination of the spectra of the two pure components. Assume that the spectrum of pure component A is $A(k)$ and the spectrum of pure component B is $B(k)$. Then, if we assume that we have a mixture of A and B and that the concentration of the components in the mixture is x_a and x_b, respectively, then the spectrum of the mixture is $x_aA(k) + x_bB(k)$. Now, we can reverse the argument. Assume we have a mixture of the components A and B, but we do not know the concentrations. If we acquire a spectrum of the mixture $E(k)$ and if we know the spectra of the pure components, could we extract the concentrations from the measured spectrum of the mixture? We could set up two curves of the type shown in Figure B.2 by using two points—spectrum $B(k)$ for $x_a = 0$, $x_b = 1$ and spectrum $A(k)$ for $x_a = 1$, $x_b = 0$. After all, we assumed the relationship is linear. Then, we pick a peak in the spectrum of A that is either not present in the spectrum of B or has a height as different from the peak in B as possible. Then, we plot the peak height from spectrum $B(k)$ on the graph such as shown in Figure B.2 for $x_a = 0$ and plot the peak height from spectrum $A(k)$ for $x_a = 1$and draw a straight line between the two points. This becomes equivalent of the procedure that we outlined above that resulted in Figure B.2. However, we are now interested in an approach that would allow us to use the full spectrum $E(k)$ to extract x_a and x_b. To see how it works, let us write

$$M(x_a, x_b) = \int \{E(k) - [x_aA(k) + x_bB(k)]\}^2 \, dk. \tag{B.5}$$

Let us look closely at an interesting feature of the expression (Eq. B.5). Since the expression under the integral is squared, it is positive for every wave number. So, the integral is positive for every value of x_a and x_b. The only way that M can ever be zero is if

$$E(k) = x_a A(k) + x_b B(k). \tag{B.6}$$

Now, even if we assume that the spectrum of the mixture is not exactly a linear combination of the spectra of the pure components, it would seem quite plausible that M is minimized for the actual concentrations of the components in the mixture. This means that

$$\begin{aligned} \frac{\partial M(x_a, x_b)}{\partial x_a} &= 0 \\ \frac{\partial M(x_a, x_b)}{\partial x_b} &= 0. \end{aligned} \tag{B.7}$$

Carrying out the implied operations in Equations B.5 and B.7 yields a system of two linear equations with two unknowns:

$$\begin{aligned} x_a M_{aa} + x_b M_{ab} &= M_{ea} \\ x_a M_{ab} + x_b M_{bb} &= M_{eb}, \end{aligned} \tag{B.8}$$

where

$$M_{aa} = \int [A(k)]^2 dk,$$

$$M_{bb} = \int [B(k)]^2 dk,$$

$$M_{ab} = \int A(k) B(k) dk,$$

$$M_{ea} = \int A(k) E(k) dk,$$

and

$$M_{eb} = \int B(k) E(k) dk.$$

The solution of the system (Eq. B.8) is easy to find by direct calculation:

$$\begin{aligned} x_a &= \frac{M_{bb} M_{ea} - M_{ab} M_{eb}}{M_{aa} M_{bb} - M_{ab}^2} \\ x_b &= \frac{M_{aa} M_{eb} - M_{ab} M_{ea}}{M_{aa} M_{bb} - M_{ab}^2}. \end{aligned} \tag{B.9}$$

The result (Eq. B.9) demonstrates how the entire spectra of the pure components and of the mixture can be used to extract the unknown concentrations. Since all available information is used, no information is lost.

We have used x_a and x_b as independent variables. However, since the two components are combined in a new whole, would it not be reasonable to expect that they are not independent, but that x_a and x_b add to one? It is certainly true that the two concentrations must add into a number that is close to one, but they do not have to add exactly to one. In some cases, adding one component into another does not increase the volume of the solution. In other words, the sum of the two concentrations is larger than one. The two-component picture just presented is easily extended to a multiple component system.

The method above is not applicable if the spectra of the pure components are not available. Also, sometimes, the mixing of two components influences the spectra of the individual components. Peaks may shift positions, strengths, and widths; they may split into two, and so on. These spectral distortions are usually referred to as matrix effects. For instance, imagine the previously mentioned spectra of wine, which are essentially alcohol in water. To analyze a spectrum of, say, 15% alcohol in water, it would be preferable to use spectra of say 10% and 20% alcohol in water instead of using spectra of pure water and alcohol. That would take into account most of the matrix effects already present in the spectra of the two reference mixtures, $A_1(k)$ and $A_2(k)$. To see how it works, imagine that we have two spectra of the reference solutions as

$$\begin{aligned} A(k) &= a_1 W(k) + b_1 V(k) \\ B(k) &= a_2 W(k) + b_2 V(k), \end{aligned} \tag{B.10}$$

where $W(k)$ and $V(k)$ are the (unknown) spectra of the pure components; a_1 and b_1 are their (known) concentrations in the first reference solution; and a_2 and b_2 are their (known) concentrations in the second reference solution. If we collect a spectrum $E(k)$ of the solution with unknown concentrations of the two components, we could use the above method to calculate the concentrations x_a and x_b. However, x_a and x_b are now not the the concentrations of the pure components but of the mixtures (Eq. B.10). To get the actual concentrations of the pure components c_1 and c_2, we have to calculate

$$\begin{aligned} c_1 &= x_a a_1 + x_b a_2 \\ c_2 &= x_a b_1 + x_b b_2. \end{aligned} \tag{B.11}$$

Finally, there is a powerful statistical analysis method called principal component decomposition. This method has many variations and this is not the place to go into details, but the basic ideas behind the method can be illustrated with an example. Imagine you want to analyze the single reflection ATR spectra of cooking oils such as sunflower, olive, corn, and peanut. You amass a huge library of individual oils and start studying the spectra. The first thing you notice is that the spectra of all oils are very similar. There is obviously a good reason why all these substances are called cooking oils. So you decide that most of the information contained in the spectrum of each oil relates to the common oiliness of the sample, not to the particular type of oil. Thus, a good way to start manipulating the spectra of oils would be to find the average value of all the spectra and then to subtract the average from the spectrum of each of the oils. The features related to the "oiliness" of the sample are thus subtracted from each spectrum, and the residual spectra contain the features that are specific to each particular type of oil. The step described is very obvious. The problem is how to keep on doing the same type of thing. A beautiful mathematical formalism can be brought in to help. The formalism enables us to conceive the spectra as vectors in some imaginary space. Spectra are essentially functions of the wave number and can be added together and can also be multiplied by a constant. To make them into vectors, we need a way to calculate the scalar product of the two spectra. Imagine two spectra, $S_1(k)$ and $S_2(k)$. Then, the scalar product of these two spectra can be defined as

$$\boldsymbol{S_1} \cdot \boldsymbol{S_2} = \int S_1(k) S_2(k) dk, \tag{B.12}$$

where the integration extends over the spectral region of interest. Once this is defined, it is easy to find the norm (length) of a spectrum as the square root of the scalar product of the spectrum with itself. In general, a scalar product of two vectors, **a** and **b**, is

$$\mathbf{a} \cdot \mathbf{b} = ab\cos(\alpha)$$

where a and b are the lengths of the respective vectors and α is the angle between them.

Once we can calculate the length of a spectrum, we can divide a spectrum with its length and convert a spectrum into a unit vector (a vector of unit length). Or we can divide the scalar product of two spectra by the product of their lengths, and the result is the cosine of the angle between them. These are mind-boggling concepts. They let us calculate the length of each spectrum and the angle between any two spectra. So if the angle between two spectra is 90°, the two spectra are perpendicular to each other. If the angle is 0°, the spectra are parallel. The entire machinery of vector analysis can be used to manipulate spectra. If two spectra, **a** and **b**, are not parallel (we then call them linearly independent), we can decompose **b** into a component $\mathbf{b}^{\parallel}$ parallel to **a** and a component $\mathbf{b}^{\perp}$ perpendicular to **a** as follows:

$$\mathbf{b}^{\parallel} = \mathbf{a}(\mathbf{ab}) / \mathrm{a}^2, \mathbf{b}^{\perp} = \mathbf{b} - \mathbf{b}^{\parallel}. \tag{B.13}$$

Thus, $\mathbf{e}_1 = \mathbf{a}/|\mathbf{a}|$ and $\mathbf{e}_2 = (\mathbf{b} - \mathbf{b}^{\parallel})/|\mathbf{b} - \mathbf{b}^{\parallel}|$ are two mutually perpendicular unit vectors. The method shown above can be continued to build an orthonormal base of unit vectors for any number of linearly independent vectors.

Coming back to our cooking oil example, we see that we have subtracted common oiliness from all the spectra. Let us assume that we have N spectra of oils. We can now normalize the average spectrum and call it $\mathbf{e}_0$. Each of the N spectra $\mathrm{S}_n(k)$ could be separated into a component parallel to $\mathbf{e}_0$ and a component perpendicular to $\mathbf{e}_0$ as follows:

$$\mathbf{S}_n = \mathbf{e}_0(\mathbf{S}_n \cdot \mathbf{e}_0) + (\mathbf{S}_n - \mathbf{e}_0(\mathbf{S}_n \cdot \mathbf{e}_0)) = \mathbf{S}_n^{\perp} + \mathbf{S}_n^{\parallel}. \tag{B.14}$$

We can then drop the components parallel to $\mathbf{e}_0$ since they do not carry independent information. Next, we can find the lengths of the remaining spectra and pick the largest one, normalize it, and call it $\mathbf{e}_1$. Again, we can subtract from each of the remaining spectra the component parallel to $\mathbf{e}_1$, find lengths of the remaining spectra, pick the largest one, and so on. Continuing along this way, the process will terminate in M steps, where M is smaller or equal to N. Therefore, we end up with M orthogonal unit vectors. These vectors form an orthonormal base of the space spanned by the original oil spectra. That means that we can decompose each spectrum as

$$\mathbf{S}_n = \sum c_{nk} \mathbf{e}_k,$$

where the coefficients c_{nk} are simply the scalar products $\mathbf{S}_n \cdot \mathbf{e}_k$. The base vectors are called the principal components. This, at first sight, may not seem as much of an accomplishment, but think of what has been achieved. The original spectra would typically be acquired at a resolution of $4\,\mathrm{cm}^{-1}$ in the spectral range from 400 to $4000\,\mathrm{cm}^{-1}$. This is 3600/4 = 900 independent data points in each spectrum. If we wanted to represent a spectrum as a point in space where each wave number data point is plotted on its own independent axis, we would need a 900-dimensional space. At a higher resolution and/or for a wider spectral range, we would need even more dimensions. This is a bewilderingly huge space. The procedure just described reduces this space to a much smaller space consisting of only those dimensions in the spectral space that contribute truly independent spectral content. Then, each spectrum from the original set can be represented by a point in the spectral space subtended by the orthonormal base of unit vectors $\mathbf{e}_n$. The coordinates of this point are the coefficients c_{nk}. Not only can we do it for any of the spectra from the initial set, but we can do it with any mixture of these spectra as well. We can also decompose each new spectrum of oil that we come across using the same base. If we discover that there is a significant residual after the decomposition of the new spectrum (i.e., the component that is perpendicular to all the base vectors), we can normalize that residual component and add it to our base, expanding our space by one additional dimension. Note that all the prior spectra of oils had no component along this new dimension. We could also be justifiably suspicious that the particular oil may have a nonoil additive leading to the new spectral features, in which case we may choose not to expand our space to include this additional base vector. The point of the decomposition is that the dimension of this new space is usually fairly small, that is, less than 10. So, from dealing with vectors having 900 or so independent components, we end up with vectors having 10 or so independent components. It is then much easier to analyze experimental data even for characteristics of a sample that are not directly definable in terms of sample composition such as unsaturation of oil, or acidity of oil, or the octane number of gasoline. We first "train" the model by a set of known standards. This is done by finding the coordinates of each standard in the base of principal components and then finding how the trait that is known for standards correlates with each of the coordinates of the standard. It is then possible to decompose a spectrum of an unknown sample into principal components and, from the coordinates of the spectrum and the established correlations with the trait of interest, to predict the trait for the unknown sample.

Another use of the principal component decomposition comes from its ability to reveal clustering in a set of samples. In our cooking oil example, we could, for instance, study the subset of all olive oils in the set. Some of those oils may be from, say, Spain and some from Italy. We could graph the entire set in the space spanned by its principal components. It is impossible for us to visualize a space with more than three dimensions, but we can project all the points in the set onto a two-dimensional plane in this multidimensional space. It is conceivable that there is a plane such that if the points are projected into this plane, the set would resolve into two clearly distinguished clusters—one consisting of Spanish and another consisting of Italian oils. Then, by analyzing an unknown olive oil, one can easily examine how that oil fits into this clustering and be able to say something about the country of origin of the unknown oil. A similar clustering scheme can be used to analyze complex biological tissues (cancerous vs. normal cells), crude oils with respect to their origin, wines, and so on.

Note that principal components are not spectra. They must have both positive and negative values in order to be mutually orthogonal. Two real spectra, having only nonnegative values, can never be mutually orthogonal. The negative values make principal components appear somewhat unphysical. Nevertheless, the method is extremely powerful. Many variations of this technique are currently in use, and new variants are being developed all the time.

Let us finish by demonstrating how a huge spectral space (900 independent values in the example above) can be reduced to a space having only a small number of dimensions. Imagine that we prepare a number of mixtures of the same 10 ingredients mixed in different concentrations. Clearly, the spectrum of any possible mixture is constrained by having only those spectral features that are present in the spectra of the 10 pure components. For instance, if none of the 10 ingredients has a peak at a particular wave number, none of the mixtures will have a peak at that wave number. It is clear that there are only 10 independent parameters in the space of all possible mixtures and those are the concentrations of the components in the mixture. So, a space containing the spectra of all possible mixtures of the 10 ingredients contains only 10 linearly independent spectra. The above described procedure of building an orthonormal base of principal components would, in this case, end at 10 principal components. Each additional mixture spectrum is expressible as a linear combination of these 10 principal components. The original 900-dimensional space of all possible spectra reduces to a 10-dimensional subspace.

INDEX

Internal Reflection and ATR Spectroscopy, First Edition. Milan Milosevic.

CHEMICAL ANALYSIS

A SERIES OF MONOGRAPHS ON ANALYTICAL CHEMISTRY AND ITS APPLICATIONS

Series Editor
MARK F. VITHA

Vol. 1 **The Analytical Chemistry of Industrial Poisons, Hazards, and Solvents**. *Second Edition.* By the late Morris B. Jacobs
Vol. 2 **Chromatographic Adsorption Analysis**. By Harold H. Strain (*out of print*)
Vol. 3 **Photometric Determination of Traces of Metals**. *Fourth Edition*
Part I: General Aspects. By E. B. Sandell and Hiroshi Onishi
Part IIA: Individual Metals, Aluminum to Lithium. By Hiroshi Onishi
Part IIB: Individual Metals, Magnesium to Zirconium. By Hiroshi Onishi
Vol. 4 **Organic Reagents Used in Gravimetric and Volumetric Analysis**. By John F. Flagg (*out of print*)
Vol. 5 **Aquametry: A Treatise on Methods for the Determination of Water**. *Second Edition* (*in three parts*). By John Mitchell, Jr. and Donald Milton Smith
Vol. 6 **Analysis of Insecticides and Acaricides**. By Francis A. Gunther and Roger C. Blinn (*out of print*)
Vol. 7 **Chemical Analysis of Industrial Solvents**. By the late Morris B. Jacobs and Leopold Schetlan
Vol. 8 **Colorimetric Determination of Nonmetals**. *Second Edition.* Edited by the late David F. Boltz and James A. Howell
Vol. 9 **Analytical Chemistry of Titanium Metals and Compounds**. By Maurice Codell
Vol. 10 **The Chemical Analysis of Air Pollutants**. By the late Morris B. Jacobs
Vol. 11 **X-Ray Spectrochemical Analysis**. *Second Edition.* By L. S. Birks
Vol. 12 **Systematic Analysis of Surface-Active Agents**. *Second Edition.* By Milton J. Rosen and Henry A. Goldsmith
Vol. 13 **Alternating Current Polarography and Tensammetry**. By B. Breyer and H.H. Bauer
Vol. 14 **Flame Photometry**. By R. Herrmann and J. Alkemade
Vol. 15 **The Titration of Organic Compounds** (*in two parts*). By M. R. F. Ashworth
Vol. 16 **Complexation in Analytical Chemistry: A Guide for the Critical Selection of Analytical Methods Based on Complexation Reactions**. By the late Anders Ringbom
Vol. 17 **Electron Probe Microanalysis**. *Second Edition.* By L. S. Birks
Vol. 18 **Organic Complexing Reagents: Structure, Behavior, and Application to Inorganic Analysis**. By D. D. Perrin
Vol. 19 **Thermal Analysis**. *Third Edition.* By Wesley Wm. Wendlandt
Vol. 20 **Amperometric Titrations**. By John T. Stock
Vol. 21 **Reflctance Spectroscopy**. By Wesley Wm. Wendlandt and Harry G. Hecht
Vol. 22 **The Analytical Toxicology of Industrial Inorganic Poisons**. By the late Morris B. Jacobs
Vol. 23 **The Formation and Properties of Precipitates**. By Alan G. Walton
Vol. 24 **Kinetics in Analytical Chemistry**. By Harry B. Mark, Jr. and Garry A. Rechnitz
Vol. 25 **Atomic Absorption Spectroscopy**. *Second Edition.* By Morris Slavin
Vol. 26 **Characterization of Organometallic Compounds** (*in two parts*). Edited by Minoru Tsutsui
Vol. 27 **Rock and Mineral Analysis**. *Second Edition.* By Wesley M. Johnson and John A. Maxwell
Vol. 28 **The Analytical Chemistry of Nitrogen and Its Compounds** (*in two parts*). Edited by C. A. Streuli and Philip R. Averell
Vol. 29 **The Analytical Chemistry of Sulfur and Its Compounds** (*in three parts*). By J. H. Karchmer
Vol. 30 **Ultramicro Elemental Analysis**. By Güther Toölg
Vol. 31 **Photometric Organic Analysis** (*in two parts*). By Eugene Sawicki
Vol. 32 **Determination of Organic Compounds: Methods and Procedures**. By Frederick T. Weiss
Vol. 33 **Masking and Demasking of Chemical Reactions**. By D. D. Perrin

Vol. 34 **Neutron Activation Analysis**. By D. De Soete, R. Gijbels, and J. Hoste
Vol. 35 **Laser Raman Spectroscopy**. By Marvin C. Tobin
Vol. 36 **Emission Spectrochemical Analysis**. By Morris Slavin
Vol. 37 **Analytical Chemistry of Phosphorus Compounds**. Edited by M. Halmann
Vol. 38 **Luminescence Spectrometry in Analytical Chemistry**. By Mark F. Vitha, S. G. Schulman, and T. C. O'Haver
Vol. 39. **Activation Analysis with Neutron Generators**. By Sam S. Nargolwalla and Edwin P. Przybylowicz
Vol. 40 **Determination of Gaseous Elements in Metals**. Edited by Lynn L. Lewis, Laben M. Melnick, and Ben D. Holt
Vol. 41 **Analysis of Silicones**. Edited by A. Lee Smith
Vol. 42 **Foundations of Ultracentrifugal Analysis**. By H. Fujita
Vol. 43 **Chemical Infrared Fourier Transform Spectroscopy**. By Peter R. Griffiths
Vol. 44 **Microscale Manipulations in Chemistry**. By T. S. Ma and V. Horak
Vol. 45 **Thermometric Titrations**. By J. Barthel
Vol. 46 **Trace Analysis: Spectroscopic Methods for Elements**. Edited by Mark F. Vitha
Vol. 47 **Contamination Control in Trace Element Analysis**. By Morris Zief and James W. Mitchell
Vol. 48 **Analytical Applications of NMR**. By D. E. Leyden and R. H. Cox
Vol. 49 **Measurement of Dissolved Oxygen**. By Michael L. Hitchman
Vol. 50 **Analytical Laser Spectroscopy**. Edited by Nicolo Omenetto
Vol. 51 **Trace Element Analysis of Geological Materials**. By Roger D. Reeves and Robert R. Brooks
Vol. 52 **Chemical Analysis by Microwave Rotational Spectroscopy**. By Ravi Varma and Lawrence W. Hrubesh
Vol. 53 **Information Theory as Applied to Chemical Analysis**. By Karl Eckschlager and Vladimir Stepanek
Vol. 54 **Applied Infrared Spectroscopy: Fundamentals, Techniques, and Analytical Problemsolving**. By A. Lee Smith
Vol. 55 **Archaeological Chemistry**. By Zvi Goffer
Vol. 56 **Immobilized Enzymes in Analytical and Clinical Chemistry**. By P. W. Carr and L. D. Bowers
Vol. 57 **Photoacoustics and Photoacoustic Spectroscopy**. By Allan Rosencwaig
Vol. 58 **Analysis of Pesticide Residues**. Edited by H. Anson Moye
Vol. 59 **Affity Chromatography**. By William H. Scouten
Vol. 60 **Quality Control in Analytical Chemistry**. *Second Edition*. By G. Kateman and L. Buydens
Vol. 61 **Direct Characterization of Fineparticles**. By Brian H. Kaye
Vol. 62 **Flow Injection Analysis**. By J. Ruzicka and E. H. Hansen
Vol. 63 **Applied Electron Spectroscopy for Chemical Analysis**. Edited by Hassan Windawi and Floyd Ho
Vol. 64 **Analytical Aspects of Environmental Chemistry**. Edited by David F. S. Natusch and Philip K. Hopke
Vol. 65 **The Interpretation of Analytical Chemical Data by the Use of Cluster Analysis**. By D. Luc Massart and Leonard Kaufman
Vol. 66 **Solid Phase Biochemistry: Analytical and Synthetic Aspects**. Edited by William H. Scouten
Vol. 67 **An Introduction to Photoelectron Spectroscopy**. By Pradip K. Ghosh
Vol. 68 **Room Temperature Phosphorimetry for Chemical Analysis**. By Tuan Vo-Dinh
Vol. 69 **Potentiometry and Potentiometric Titrations**. By E. P. Serjeant
Vol. 70 **Design and Application of Process Analyzer Systems**. By Paul E. Mix
Vol. 71 **Analysis of Organic and Biological Surfaces**. Edited by Patrick Echlin
Vol. 72 **Small Bore Liquid Chromatography Columns: Their Properties and Uses**. Edited by Raymond P. W. Scott
Vol. 73 **Modern Methods of Particle Size Analysis**. Edited by Howard G. Barth
Vol. 74 **Auger Electron Spectroscopy**. By Michael Thompson, M. D. Baker, Alec Christie, and J. F. Tyson
Vol. 75 **Spot Test Analysis: Clinical, Environmental, Forensic and Geochemical Applications**. By Ervin Jungreis
Vol. 76 **Receptor Modeling in Environmental Chemistry**. By Philip K. Hopke

Vol. 77 **Molecular Luminescence Spectroscopy: Methods and Applications** (*in three parts*). Edited by Stephen G. Schulman
Vol. 78 **Inorganic Chromatographic Analysis**. Edited by John C. MacDonald
Vol. 79 **Analytical Solution Calorimetry**. Edited by J. K. Grime
Vol. 80 **Selected Methods of Trace Metal Analysis: Biological and Environmental Samples**. By Jon C. VanLoon
Vol. 81 **The Analysis of Extraterrestrial Materials**. By Isidore Adler
Vol. 82 **Chemometrics**. By Muhammad A. Sharaf, Deborah L. Illman, and Bruce R. Kowalski
Vol. 83 **Fourier Transform Infrared Spectrometry**. By Peter R. Griffiths and James A. de Haseth
Vol. 84 **Trace Analysis: Spectroscopic Methods for Molecules**. Edited by Gary Christian and James B. Callis
Vol. 85 **Ultratrace Analysis of Pharmaceuticals and Other Compounds of Interest**. Edited by S. Ahuja
Vol. 86 **Secondary Ion Mass Spectrometry: Basic Concepts, Instrumental Aspects, Applications and Trends**. By A. Benninghoven, F. G. Rüenauer, and H. W. Werner
Vol. 87 **Analytical Applications of Lasers**. Edited by Edward H. Piepmeier
Vol. 88 **Applied Geochemical Analysis**. By C. O. Ingamells and F. F. Pitard
Vol. 89 **Detectors for Liquid Chromatography**. Edited by Edward S. Yeung
Vol. 90 **Inductively Coupled Plasma Emission Spectroscopy: Part 1: Methodology, Instrumentation, and Performance; Part II: Applications and Fundamentals**. Edited by J. M. Boumans
Vol. 91 **Applications of New Mass Spectrometry Techniques in Pesticide Chemistry**. Edited by Joseph Rosen
Vol. 92 **X-Ray Absorption: Principles, Applications, Techniques of EXAFS, SEXAFS, and XANES**. Edited by D. C. Konnigsberger
Vol. 93 **Quantitative Structure-Chromatographic Retention Relationships**. By Roman Kaliszan
Vol. 94 **Laser Remote Chemical Analysis**. Edited by Raymond M. Measures
Vol. 95 **Inorganic Mass Spectrometry**. Edited by F. Adams, R. Gijbels, and R. Van Grieken
Vol. 96 **Kinetic Aspects of Analytical Chemistry**. By Horacio A. Mottola
Vol. 97 **Two-Dimensional NMR Spectroscopy**. By Jan Schraml and Jon M. Bellama
Vol. 98 **High Performance Liquid Chromatography**. Edited by Phyllis R. Brown and Richard A. Hartwick
Vol. 99 **X-Ray Fluorescence Spectrometry**. By Ron Jenkins
Vol. 100 **Analytical Aspects of Drug Testing**. Edited by Dale G. Deustch
Vol. 101 **Chemical Analysis of Polycyclic Aromatic Compounds**. Edited by Tuan Vo-Dinh
Vol. 102 **Quadrupole Storage Mass Spectrometry**. By Raymond E. March and Richard J. Hughes (*out of print: see Vol. 165)*
Vol. 103 **Determination of Molecular Weight**. Edited by Anthony R. Cooper
Vol. 104 **Selectivity and Detectability Optimization in HPLC**. By Satinder Ahuja
Vol. 105 **Laser Microanalysis**. By Lieselotte Moenke-Blankenburg
Vol. 106 **Clinical Chemistry**. Edited by E. Howard Taylor
Vol. 107 **Multielement Detection Systems for Spectrochemical Analysis**. By Kenneth W. Busch and Marianna A. Busch
Vol. 108 **Planar Chromatography in the Life Sciences**. Edited by Joseph C. Touchstone
Vol. 109 **Fluorometric Analysis in Biomedical Chemistry: Trends and Techniques Including HPLC Applications**. By Norio Ichinose, George Schwedt, Frank Michael Schnepel, and Kyoko Adochi
Vol. 110 **An Introduction to Laboratory Automation**. By Victor Cerdá and Guillermo Ramis
Vol. 111 **Gas Chromatography: Biochemical, Biomedical, and Clinical Applications**. Edited by Ray E. Clement
Vol. 112 **The Analytical Chemistry of Silicones**. Edited by A. Lee Smith
Vol. 113 **Modern Methods of Polymer Characterization**. Edited by Howard G. Barth and Jimmy W. Mays
Vol. 114 **Analytical Raman Spectroscopy**. Edited by Jeanette Graselli and Bernard J. Bulkin
Vol. 115 **Trace and Ultratrace Analysis by HPLC**. By Satinder Ahuja
Vol. 116 **Radiochemistry and Nuclear Methods of Analysis**. By William D. Ehmann and Diane E. Vance

Vol. 117 **Applications of Fluorescence in Immunoassays**. By Ilkka Hemmila
Vol. 118 **Principles and Practice of Spectroscopic Calibration**. By Howard Mark
Vol. 119 **Activation Spectrometry in Chemical Analysis**. By S. J. Parry
Vol. 120 **Remote Sensing by Fourier Transform Spectrometry**. By Reinhard Beer
Vol. 121 **Detectors for Capillary Chromatography**. Edited by Herbert H. Hill and Dennis McMinn
Vol. 122 **Photochemical Vapor Deposition**. By J. G. Eden
Vol. 123 **Statistical Methods in Analytical Chemistry**. By Peter C. Meier and Richard Züd
Vol. 124 **Laser Ionization Mass Analysis**. Edited by Akos Vertes, Renaat Gijbels, and Fred Adams
Vol. 125 **Physics and Chemistry of Solid State Sensor Devices**. By Andreas Mandelis and Constantinos Christofides
Vol. 126 **Electroanalytical Stripping Methods**. By Khjena Z. Brainina and E. Neyman
Vol. 127 **Air Monitoring by Spectroscopic Techniques**. Edited by Markus W. Sigrist
Vol. 128 **Information Theory in Analytical Chemistry**. By Karel Eckschlager and Klaus Danzer
Vol. 129 **Flame Chemiluminescence Analysis by Molecular Emission Cavity Detection**. Edited by David Stiles, Anthony Calokerinos, and Alan Townshend
Vol. 130 **Hydride Generation Atomic Absorption Spectrometry**. Edited by Jiri Dedina and Dimiter L. Tsalev
Vol. 131 **Selective Detectors: Environmental, Industrial, and Biomedical Applications**. Edited by Robert E. Sievers
Vol. 132 **High-Speed Countercurrent Chromatography**. Edited by Yoichiro Ito and Walter D. Conway
Vol. 133 **Particle-Induced X-Ray Emission Spectrometry**. By Sven A. E. Johansson, John L. Campbell, and Klas G. Malmqvist
Vol. 134 **Photothermal Spectroscopy Methods for Chemical Analysis**. By Stephen E. Bialkowski
Vol. 135 **Element Speciation in Bioinorganic Chemistry**. Edited by Sergio Caroli
Vol. 136 **Laser-Enhanced Ionization Spectrometry**. Edited by John C. Travis and Gregory C. Turk
Vol. 137 **Fluorescence Imaging Spectroscopy and Microscopy**. Edited by Xue Feng Wang and Brian Herman
Vol. 138 **Introduction to X-Ray Powder Diffractometry**. By Ron Jenkins and Robert L. Snyder
Vol. 139 **Modern Techniques in Electroanalysis**. Edited by Petr Vanýek
Vol. 140 **Total-Reflction X-Ray Fluorescence Analysis**. By Reinhold Klockenkamper
Vol. 141 **Spot Test Analysis: Clinical, Environmental, Forensic, and Geochemical Applications**. *Second Edition.* By Ervin Jungreis
Vol. 142 **The Impact of Stereochemistry on Drug Development and Use**. Edited by Hassan Y. Aboul-Enein and Irving W. Wainer
Vol. 143 **Macrocyclic Compounds in Analytical Chemistry**. Edited by Yury A. Zolotov
Vol. 144 **Surface-Launched Acoustic Wave Sensors: Chemical Sensing and Thin-Film Characterization**. By Michael Thompson and David Stone
Vol. 145 **Modern Isotope Ratio Mass Spectrometry**. Edited by T. J. Platzner
Vol. 146 **High Performance Capillary Electrophoresis: Theory, Techniques, and Applications**. Edited by Morteza G. Khaledi
Vol. 147 **Solid Phase Extraction: Principles and Practice**. By E. M. Thurman
Vol. 148 **Commercial Biosensors: Applications to Clinical, Bioprocess and Environmental Samples**. Edited by Graham Ramsay
Vol. 149 **A Practical Guide to Graphite Furnace Atomic Absorption Spectrometry**. By David J. Butcher and Joseph Sneddon
Vol. 150 **Principles of Chemical and Biological Sensors**. Edited by Dermot Diamond
Vol. 151 **Pesticide Residue in Foods: Methods, Technologies, and Regulations**. By W. George Fong, H. Anson Moye, James N. Seiber, and John P. Toth
Vol. 152 **X-Ray Fluorescence Spectrometry**. *Second Edition.* By Ron Jenkins
Vol. 153 **Statistical Methods in Analytical Chemistry**. *Second Edition.* By Peter C. Meier and Richard E. Züd
Vol. 154 **Modern Analytical Methodologies in Fat- and Water-Soluble Vitamins**. Edited by Won O. Song, Gary R. Beecher, and Ronald R. Eitenmiller
Vol. 155 **Modern Analytical Methods in Art and Archaeology**. Edited by Enrico Ciliberto and Guiseppe Spoto

Vol. 156 **Shpol'skii Spectroscopy and Other Site Selection Methods: Applications in Environmental Analysis, Bioanalytical Chemistry and Chemical Physics**. Edited by C. Gooijer, F. Ariese and J. W. Hofstraat

Vol. 157 **Raman Spectroscopy for Chemical Analysis**. By Richard L. McCreery

Vol. 158 **Large (C> = 24) Polycyclic Aromatic Hydrocarbons: Chemistry and Analysis**. By John C. Fetzer

Vol. 159 **Handbook of Petroleum Analysis**. By James G. Speight

Vol. 160 **Handbook of Petroleum Product Analysis**. By James G. Speight

Vol. 161 **Photoacoustic Infrared Spectroscopy**. By Kirk H. Michaelian

Vol. 162 **Sample Preparation Techniques in Analytical Chemistry**. Edited by Somenath Mitra

Vol. 163 **Analysis and Purification Methods in Combination Chemistry**. Edited by Bing Yan

Vol. 164 **Chemometrics: From Basics to Wavelet Transform**. By Foo-tim Chau, Yi-Zeng Liang, Junbin Gao, and Xue-guang Shao

Vol. 165 **Quadrupole Ion Trap Mass Spectrometry**. *Second Edition.* By Raymond E. March and John F. J. Todd

Vol. 166 **Handbook of Coal Analysis**. By James G. Speight

Vol. 167 **Introduction to Soil Chemistry: Analysis and Instrumentation**. By Alfred R. Conklin, Jr.

Vol. 168 **Environmental Analysis and Technology for the Refining Industry**. By James G. Speight

Vol. 169 **Identification of Microorganisms by Mass Spectrometry**. Edited by Charles L. Wilkins and Jackson O. Lay, Jr.

Vol. 170 **Archaeological Chemistry**. *Second Edition.* By Zvi Goffer

Vol. 171 **Fourier Transform Infrared Spectrometry**. *Second Edition.* By Peter R. Griffiths and James A. de Haseth

Vol. 172 **New Frontiers in Ultrasensitive Bioanalysis: Advanced Analytical Chemistry Applications in Nanobiotechnology, Single Molecule Detection, and Single Cell Analysis**. Edited by Xiao-Hong Nancy Xu

Vol. 173 **Liquid Chromatography Time-of-Flight Mass Spectrometry: Principles, Tools, and Applications for Accurate Mass Analysis.** Edited by Imma Ferrer and E. Michael Thurman

Vol. 174 **In Vivo Glucose Sensing.** Edited by David O. Cunningham and Julie A. Stenken

Vol. 175 **MALDI Mass Spectrometry for Synthetic Polymer Analysis.** By Liang Li

Vol. 176 **Internal Reflection and ATR Spectroscopy.** By Milan Milosevic